D0550309

RTC Limerick

International Construction

International Construction

Mark Mawhinney

Sustainable Cities Research Institute
University of Northumbria at Newcastle
and
School of Science and Technology
University of Teesside

Blackwell
Science

© 2001 by
Blackwell Science Ltd
Editorial Offices:
Osney Mead, Oxford OX2 0EL
25 John Street, London WC1N 2BS
23 Ainslie Place, Edinburgh EH3 6AJ
350 Main Street, Malden
   MA 02148 5018, USA
54 University Street, Carlton
   Victoria 3053, Australia
10, rue Casimir Delavigne
   75006 Paris, France

Other Editorial Offices:

Blackwell Wissenschafts-Verlag GmbH
Kurfürstendamm 57
10707 Berlin, Germany

Blackwell Science KK
MG Kodenmacho Building
7–10 Kodenmacho Nihombashi
Chuo-ku, Tokyo 104, Japan

Iowa State University Press
A Blackwell Science Company
2121 S. State Avenue
Ames, Iowa 50014-8300, USA

First published 2001

Set in 10/13pt Palatino
by DP Photosetting, Aylesbury, Bucks
Printed and bound in Great Britain by
MPG Books Ltd, Bodmin, Cornwall

The Blackwell Science logo is a
trade mark of Blackwell Science Ltd,
registered at the United Kingdom
Trade Marks Registry

DISTRIBUTORS
   Marston Book Services Ltd
   PO Box 269
   Abingdon
   Oxon OX14 4YN
   (Orders: Tel: 01235 465500
              Fax: 01235 465555)

USA
   Blackwell Science, Inc.
   Commerce Place
   350 Main Street
   Malden, MA 02148 5018
   (Orders: Tel: 800 759 6102
                   781 388 8250
              Fax: 781 388 8255)

Canada
   Login Brothers Book Company
   324 Saulteaux Crescent
   Winnipeg, Manitoba R3J 3T2
   (Orders: Tel: 204 837-2987
              Fax: 204 837-3116)

Australia
   Blackwell Science Pty Ltd
   54 University Street
   Carlton, Victoria 3053
   (Orders: Tel: 03 9347 0300
              Fax: 03 9347 5001)

A catalogue record for this title
is available from the British Library

ISBN 0-632-05853-6

Library of Congress
Cataloging-in-Publication Data

Mawhinney, Mark.
   International construction/Mark Mawhinney.
      p. cm.
   Includes bibliographical references and index.
   ISBN 0-632-05853-6
   1. Construction industry.   2. Construction
industry – Management.   I. Title.

HD9715.A2 M346 2001
624'.068 – dc21   2001029524

For further information on
Blackwell Science, visit our website:
www.blackwell-science.com

Contents

Abbreviations

ACE	Association of Consulting Engineers
ASEAN	Association of South East Asian Nations
BCB	British Consultants Bureau
BCSA	British Constructional Steel Association
BERI	Business Environment Risk Intelligence
BOO	Build, own and operate
BOOT	Build, own, operate and transfer
CIA	Central Intelligence Agency (USA)
CIRIA	Construction Industry Research and Information Association
CPA	Construction Products Association
CTRL	Channel Tunnel Rail Link
DBFO	Design, build, finance and operate
DCF	Discounted cash-flow
DCMF	Design, construct, maintain and finance
DETR	Department of the Environment, Transport and the Regions (UK)
DFID	Department for International Development (UK)
DoE	Department of the Environment (UK)
DTI	Department of Trade and Industry (UK)
EBRD	European Bank for Reconstruction and Development
ECGD	Export Credit Guarantee Department
ECI	European Construction Institute
ECU	European currency unit
EFCA	European Federation of Engineering Consultancy Association
EGCI	Export Group for Constructional Industries (UK trade association for contractors)
EIB	European Investment Bank
EIC	European International Contractors
ENR	*Engineering-News Record* (American magazine)
EU	European Union
EXIM	Export Import Bank (USA)
FCO	Foreign and Commonwealth Office

FEACO	European Federation of Management Consulting Associations
GCAO	General Contractors Association of Osaka
GDP	Gross domestic product
GNP	Gross national product
IMD	Institute for Management Development (International)
IMF	International Monetary Fund
IT	Information technology
JCCA	Association of Japanese Consulting Engineers
JEXIM	Japanese Export-Import Bank
JFCC	Japanese Federation of Construction Contractors
KOZAI	Japan Iron and Steel Exporters Association
M&A	Merger and acquisition
M&E	Mechanical and electrical engineering
MBA	Master of Business Administration
MERCOSUR	Mercado Comun del Sur (Southern (American) Common Market)
MITI	Ministry of Industry and Trade (Japan)
NAFTA	North American Free Trade Agreement
NCE	*New Civil Engineer*
NPV	Net present-day values
OCAJI	Overseas Construction Association of Japan Incorporated
OECD	Organisation for Economic Co-operation and Development
PARTS	Players, Added value, Rules, Tactics, Scope
PEST	Political, Economic, Social, Technical
PESTLE	Political, Economic, Social, Technical, Legal, Environmental
PFI	Private Finance Initiative
PPA	Power purchase agreement
R&D	Research and development
RICE	Research Institute of Construction and Economy
SMART	Suitable, Measurable, Achievable, Realistic, Timely
SWOT	Strengths, Weaknesses, Opportunities, Threats
TENS	Trans-European Network Scheme
USP	Unique selling points
WEO	World Economic Outlook

Preface

A few years ago I took on the apparently straightforward task of developing an international construction module for the construction degree course at Teesside University. Experience in an international business followed by work in government assisting other companies had given me the false impression that plenty of information was available. However, it quickly became obvious that much of the information is anecdotal, confidential or very specific to one particular case.

Students need standard texts which will help them build up a picture of the market. In the commercial world that 'standard text' is provided through the guidance of experienced colleagues and through the experience of competition, something not available to students. The information made available often requires skills in matching the pieces together to complete a jigsaw.

It was clear that there are few if any standard texts in this field and there is a gap in the market for undergraduates. In fact downsizing and the rush to off-load expensive more experienced staff in recent years may have left the commercial world with gaps to fill.

With the globalisation of work moving at such a frenetic pace it was pointless trying to develop a text which described the world as it stood still on one day. We have for example seen South East Asia move from star to outcast and back to star status in the space of a few years. Instead, what is needed is a book which will provide a basic understanding of the global construction market, an explanation of what information is required and how to retrieve it, and will also look at some simple tools for analysing situations in this field.

The text is based on a current course designed for final year undergraduates. However, the text is suitable for a wide range of individuals starting careers in construction whether they be in government support organisations or in the private sector as consultants or contractors.

The lure of a career in international rather than purely domestic construction is a particularly strong one. There are few individuals I know who do not lust after a period abroad. My own experience has convinced me that the rewards of the experience often match the perception and I hope that this text will start a few careers by dispelling any mystery and fear. It is, after all, usually similar to the domestic market but with a few complications!

Finally I would like to thank Penta Ocean Construction Ltd, Japan, for their immense help in putting this book together. In particular I would like to thank Mr Kuroki, Mr Nakayama and Mr Muramoto for tracking down difficult information and the photographs which add a little light relief to the text.

Mark Mawhinney

1 Introduction

1.1 Introduction

It is important at the outset to outline the objectives of the book, so that any reader can quickly decide the value of the text and adapt it to their particular needs. This text has been written specifically for an audience of knowledgeable construction graduates who are nevertheless newcomers to the international construction market. It is not intended to cover all aspects of the subject but it should provide an overall framework from which students or people at work can start to research the subject further.

The approach adopted follows that of MBA based course materials, with text or theory supported by short case studies as illustration. Larger case studies are employed on specific topics of importance. It is likely therefore that secondary audiences, MBA students and graduates, will be interested in the case study material (particularly in Chapters 6 to 8).

International construction as a subject needs to incorporate study in economics, socio-cultural factors and anecdotal evidence. Engineering clearly plays a part and, for example, climate, earthquakes and ground types cause many differences in international practice. However, this has been relegated to a minor supporting role in the text so that management issues associated with international aspects of construction can be highlighted.

Like much of management as a theory there is little black and white or right and wrong, and to engineering graduates, who have a healthy desire to find the ultimate solution, it can be a difficult process to seek only solutions which are agreed but not optimal.

The book has three clear objectives:

(1) *To provide a general understanding of the international construction market as a background to both business and project level work*

Work on international projects needs awareness of the wider picture i.e., the effect of the project on the wider company and the effect of the location on working practices and personnel all must be acknowledged if the best options are to be considered. This contrasts with well-referenced typical approaches to domestic work where project expertise is specialized, split and often disconnected from other aspects of the business.

(2) *To provide an explanation of some of the analytical tools on offer to help the decision-making process*

All of the tools highlighted in the text apply equally to domestic and international work, and good companies would use them for both situations. The tools suggested provide simple, structured approaches to the analysis required, but they are not exhaustive and there are many other tools available on the market, generally in classic texts on management.

(3) *To remove some of the myths associated with international work*

This third objective, in particular, requires much consideration. The construction industry is rife with the idea that everything about it is unique and every new project or market requires a radical new approach. The general experience in other sectors beyond construction is that the wheel does not have to be re-invented for every new international project. More importantly, the added social and cultural complexities are seldom insurmountable.

In order to traverse through the complications of the subject and provide as wide a coverage as possible the text deals with five main areas:

---

(1) Chapters 1 and 2 – *Introduction:* a broad explanation of the subject and information management.
(2) Chapter 3 – *Tools:* the generic approaches available to study the situation.

Chapter 4 – *Operation:* the various components required in the operation of the business.

Chapter 5 – *World market:* an analysis and explanation of the global and wider markets.
(3) Chapters 6, 7 and 8 – *Case studies:* a detailed look at plant, building materials, contracting and consulting.
(4) Chapter 9 – *Finance:* the financing aspect of projects.
(5) Chapter 10 – *Signposts:* the rules of thumb which apply to the market.

From a study of framework and general information requirements in the early chapters the text moves towards a concentration on case studies in the second half. Case studies range from anecdotal through market specific to company specific. Principles, systems and diversity are emphasised rather than covering every famous market or player. Some readers, for example, may be disappointed by the lack of coverage of specific markets in Asia or Eastern Europe, two popular areas in international construction circles. However, capturing the true essence of any one business location is impossible in a short text.

1.2 What is international construction?

International construction is a subject which needs a definition. It is difficult to provide a formal definition for a number of reasons which are common to all studies of construction and of international businesses.

Construction covers a wide range of products, services and activities. Official definitions, which form the basis for this text, concentrate on the directly related site and design activity and building material products. However, as Bon and Crosthwaite (2000) point out, they ignore much associated activity.

The definition of international for the purpose of this book will be where one company, resident in one country, performs work in another country. This is the simplest of definitions but raises many complications in today's global business world. The problems include identification of the nationality, for example, of a US based company working in the USA but owned by a UK company (Thorn *et al.* 1997). Officially it would be viewed as a UK company working internationally. Another case would be a company such as Christiani Nielsen, a small but well-known contractor in international projects. This was originally established in Denmark before moving to be registered in Thailand and later registering in the UK. Officially it would be viewed as a UK firm. However, in looking for work in third countries it is likely to be able to call on the help of all three governments when seeking information or support, for a variety of reasons outlined in Chapter 2.

If one were to conduct a survey of the public and ask the question 'What is international construction?', the chances are that the response would be based around big, headline infrastructure projects with an international cast of players: Hong Kong Airport, the Channel Tunnel or the controversial Three Gorges Dam in China.

Case Study 1.1: Hong Kong Airport (Thomson and Oakervee 1998; Wheeler 1998)

The new Hong Kong Airport would be viewed by many as a prime example of an international construction project. It is an impressive project in terms of scale and technical difficulty, requiring a 1248 hectare reclamation and airport platform, a 34 km road and rail link to the central business district and 35 000 construction workers. The total cost of US$20 bn ranks as one of the largest infrastructure developments of the world. Political tension between China and Britain, the monsoon weather of Hong Kong, the density of population around which work needed to be carried out and the huge scale of reclamation all added to the difficulty, and yet the original estimate was reduced by approximately 6% in the final out-turn cost.

The project required special labour importation legislation in order to meet the need for 35 000 workers; 225 construction contracts were agreed and signed, with 182 of these being for major work. The work by value was won by firms from the following countries: Hong Kong (23%), China (8%), Japan (26%), Britain (16%), Holland (6%), France (5%), Belgium (3%), New Zealand (3%), Australia (2%), US (2%), Spain (2%), Germany (2%) with smaller splits to Italian, South African, Austrian, Norwegian, Portuguese, Swedish and Danish firms.

At its peak the reclamation work required 18 of the world's largest dredgers working continuously 24 hours a day for 20 months.

Design of road and utility infrastructure needed to be sufficient for a working population of 45 000 (equivalent to a new town). The passenger terminal is designed to accommodate 35 million passengers per annum.

Comment

Although this is a truly international project a definition based on this type of unique project alone would be insufficient to cover the full spectrum of work which is of interest to the many players in international construction. The scale of this project, the tight deadlines and its location in the global village of Hong Kong ensured that this was probably the most international of international projects. However, the example reinforces the point that construction as an industry must be viewed within the context of the wider global economy and it is important therefore to look first at the global economy as a whole.

Before studying the link to the wider economy we must first of all look at the numerous caveats to using official statistics. The economy of one nation is an extremely complicated system and measurement of it is difficult. There is therefore a reliance on estimation in its broadest sense. The global economy is the sum of national economies and cross-national flows of goods and services. This adds to the complication, and precise measurement becomes impossible. An overview of this could become a very complicated picture and so we will simplify and then use one regional set of markets as a further example.

Case Study 1.2: The global construction market

It is debatable whether there is a global construction market, where the same players compete across the world, as in industries such as electronics, aerospace and even oil (*Economist* 1999). There are aspects of the market which have moved quickly into a global operation: the finance of private infrastructure projects, construction plant and other specialist sectors. The bulk of the market, however, appears to be a set of fragmented sub-markets with much of the work being conducted locally by local players in localised markets.

Major global organisations such as the World Bank, International Monetary Fund (IMF) and the Organisation for Economic Co-operation and Development (OECD) track global growth and the direction of progress in many sectors. They tend to split the world and categorise countries by wealth, although the terminology used can be confusing, as Table 1.1 shows.

In Table 1.1 the first column from IMF (2000) appears to be their general classification system, while the second column is specific to aid related work. There are 22 developed countries which donate aid to other less developed countries. This aid is often distributed

Case study 1.2   continued

Table 1.1   Terminology for categorising countries.

This book	IMF (2000)	IMF (2000)	Bon & Crosthwaite (2000)
Developed	Advanced	Donors (developed)	Advanced industrialised
Developing	Developing and countries in transition	Developing and countries in transition	Newly industrialised
Emerging	Countries in transition and heavily indebted poor	Low income, middle income and least developed	Least developed

to the poorest according to rough measures of wealth or development. Bon and Crosthwaite (2000), by contrast, use an industrialisation index, which will produce a slightly different list of countries.

Where possible, this book uses the simplest three categorisations of developed, developing and emerging, although information sources can make this difficult. In particular, 'countries in transition' (the countries developed out of the old Soviet bloc) often appear to be treated as a special case requiring separation.

The IMF produce a yearly report on the state of the global economy highlighting the trends for business conditions across the world, giving specific regional and national outlooks and looking specifically at the effects of multinationals and globalisation. A recent report, the World Economic Outlook (IMF 2000), suggests that the value of trade across the total global economy was worth $32 110 bn in 1999. The IMF breaks this down into three blocks, as shown in Table 1.2.

Table 1.2   Global economy.

WEO Breakdown	% Global Output	10 Year growth average
Advanced nations	50	2.8%
Developing and Countries in transition	50	5.5% −2.4%

The overall figures in Table 1.2 provide a useful comparison for further analysis of the global construction sector, and show that developed or advanced nations are major drivers of the global economy. It is, however, interesting to note the pattern during the 1990s when the advanced or developed countries, which include the USA, Japan and Countries of the EU, have grown slowly, the developing nations have grown more quickly and countries in transition have contracted.

Estimates of the size of the global construction market have been made using figures from these organisations. An estimate for 1999 (Batchelor 2000) produced a global construction market of $3600 bn, which was split between civil engineering (28%), residential (36%)and non-residential (36%). This implies that the sector represents over 10% of the global economy.

Comment

Thus, construction is an important part of the global economy, affected by and affecting all parts of the globe. Its key global players have become important beyond construction and their activity and health are tracked by a number of organisations. However, the leaders in tracking the activity are *Engineering News-Record,* a US magazine which produces annual tables of performance and regularly features articles on these players. As Tables 1.3 and 1.4 show, big contractors and consultants come from big markets and most of these are developed nations.

It is important to look at both short- and long-term trends rather than at one year in isolation. The global top 20 tables (see Tables 1.3 and 1.4) show how Japanese contractors dominated the lists in the early 1990s but have since been replaced by emerging US giants. European companies appear to have held their own on balance. These trends are studied in further detail in Chapter 5. It is important, however, to note that the definition of the biggest global contractor has changed, from the biggest overall to the one with the biggest overseas turnover, which partially explains the fall from grace of the Japanese.

Korea is the only nation represented beyond the USA, Japan and EU countries – a reflection of the importance of a base in a developed nation at the time of writing. Korea is, however, a fast developing nation which has recently achieved developed status.

Much of the interest in global tables concentrates on contractors and consultants, possibly because they are most easily identified. However, it is important to remember that there are other important sectors such as plant manufacturers, material producers and

Table 1.3 Top 20 Global Contractors 1994 and 1998.

			1994 turnover (US$000)	% International				1998 turnover (US$000)	% International
1	Shimizu Corp	Japan	17,914	5	1	Bechtel Group Inc	USA	9,771	62
2	Kajima Corp	Japan	17,765	4	2	Flour Daniel Inc	USA	9,640	55
3	Taisei Corp	Japan	16,742	4	3	Bouyges SA	France	12,517	42
4	Obayashi Corp	Japan	16,083	6	4	Skanska AB	Sweden	6,939	69
5	Mitsubishi HI Ltd	Japan	15,309	28	5	Kellogg Brown & Root	USA	6,835	70
6	Takenaka Corp	Japan	12,792	5	6	HBG	Holland	4,697	75
7	Holzmann AG	Germany	11,716	20	7	Group GTM	France	7,430	46
8	Bouyges SA	France	11,224	28	8	SGE	France	9,348	36
9	Trafalgar House	UK	9,044	75	9	Hochtief AG	Germany	6,914	48
10	Kumagai Gumi Co Ltd	Japan	8,615	10	10	Holzmann AG	Germany	7,205	45
11	GTM-Entrepose	France	7,948	41	11	Bilfinger + Berger	Germany	7,948	39
12	Toda Corp	Japan	7,096	1	12	Foster Wheeler Corp	USA	7,096	31
13	Nishimatsu Con Ltd	Japan	6,922	11	13	AMEC plc	UK	6,922	31
14	Hochtief AG	Germany	6,751	34	14	JGC Corp	Japan	6,751	30
15	Fluor Daniel Inc	USA	6,638	36	15	Technip Group	France	6,638	29
16	Bechtel Group Inc	USA	6,553	24	16	Chiyoda Corp	Japan	6,553	29
17	Kinden Corp	Japan	6,274	2	17	Hyundai Eng	Korea	6,274	29
18	Kandenko Co Ltd	Japan	6,257		18	Obayashi Corp	Japan	6,257	28
19	Sato Kogyo Co Ltd	Japan	6,144	4	19	Toyo Eng Corp	Japan	6,144	27
20	Maeda Corp	Japan	5,954	4	20	Nishimatsu Con Ltd	Japan	5,954	28

Reprinted from *ENR* (1995a; 1999) Engineering News-Record 28 August 1995 and 16 August 1999.

Table 1.4 Top 20 Global Design Consultants 1994 and 1999.

		1994 International t/over (US$000)	% of total			1994 International t/over (US$000)	% of total
1	Trafalgar House	UK	1,156	87	1	AMEC plc	UK	1,032	86
2	Nethconsult	Holland	471	63	2	Bechtel Group Inc	USA	745	35
3	Brown and Root Inc	USA	388	55	3	Foster Wheeler Corp	USA	741	76
4	ABB Lummus	USA	368	72	4	Fluor Corp	USA	690	45
5	Fugro NV	Holland	328	86	5	Nethconsult	Holland	659	100
6	SNC-Lavalin	Canada	299	50	6	Kellogg, Brown & Root	USA	634	77
7	NEDECO	Holland	255	100	7	ABB Lummus Global	USA	616	85
8	Jaakro Poyry	Finland	250	77	8	Kvaerner plc	UK	589	83
9	Heidemij	Holland	246	67	9	Fugro NV	Holland	487	89
10	Louis Berger Group	USA	234	84	10	SNC Lavalin	Canada	474	54
11	Raytheon	USA	226	32	11	Arcadis NV	Holland	391	64
12	Dar Al-Handasah	Egypt	218	100	12	Jacobs Engineering	USA	372	31
13	Bouyges SA	France	182	85	13	Jaakro Poyry Group	Finland	351	83
14	Mott MacDonald	UK	175	57	14	Dar Al-Handasah	Egypt	327	99
15	The Parsons Group	USA	170	16	15	JGC Corp	Japan	310	69
16	Phillip Holzmann	Germany	161	87	16	Louis Berger Group	USA	307	85
17	Ove Arup Partnership	UK	155	55	17	Ove Arup Partnership	UK	282	58
18	SYSTRA	France	136	89	18	Parsons Brinckerhoff	USA	270	30
19	Fluor Daniel	USA	133	17	19	Groupe EGIS	France	259	57
20	Nippon Koei	Japan	128	29	20	Black and Veatch	USA	245	37

Reprinted from *ENR* (1995b, 2000) Engineering News-Record 24 July 1995 and 17 July 2000.
Note that *ENR* uses the term design firm rather than design consultant.

financiers. Although it is important to establish the global background there are few companies who can take advantage of the whole but fragmented market. Nations still represent the purest form of market for the bigger players while smaller regions or localities are the market for the majority of firms, which are small companies.

The size of a national economy is measured in many ways although the most accepted measure is the gross domestic product (GDP), as was used in the IMF study. The GDP is defined as the sum of the added value of all activity in the economy. It is an estimate of the value of goods and services sold in an economy at a given point in time (Martin-Fagg 1996). The difficulty of calculation means that the estimate can be extremely rough and it is adjusted with time until most of the error has been accounted for, typically after 4 to 5 years. Thus, a reliable estimate of GDP can only be calculated three or four years later.

An important point is that anecdotally the construction sector represents about 5% to 15% of a national economy in value, and generally 8% to 10% appears to be the norm. Analysis by Bon and Crosthwaite (2000) using a two year global set of data has indicated a range of between 4% to 58%, which appears an extremely wide range, although with an average of 11%. This is sizeable but is seldom the main driving industry in any nation. It is therefore unlikely to compete as a priority with, for example, other manufacturing or service industries. However, there can be significant numbers of people indirectly employed in winning international work, adding to the total value to the economy, and there is significant prestige to the nation involved in winning work in an international environment. As a result there is often dedicated government support of the industry in a variety of ways, which will be illustrated in Chapter 2. As part of the wider picture, this applies equally to other sectors involved in international business, and all are tracked and supported by government in their activities.

1.3 The available markets and clients

Official statisticians tend to impose four sectors within construction (Williams 1997):

- Contracting
- Consulting

- Building material production
- Construction plant

This split can be very useful in explaining the differences in approach to the subject and the consequent apparent differences in success. These will be discussed in greater detail in Chapters 6–8, but in general it is sufficient to say that the biggest domestic players of many developed nations are, by necessity, also players in the international market.

A key group for all four sectors is that of the clients and it is important to identify these, their needs and their role within the system. The highest profile group are, of course, the multi-nationals.

Case Study 1.3: The multinational client (*Economist 1995*)

The growth of multinationals in truly global operations has been an important factor in the internationalisation of construction. The lowering of trade barriers, the movement of funds and setting up of new operations globally have created a platform for interested construction companies to follow and exploit. By 1993 it was reported that multinationals had invested US$230 bn in new foreign operations, much of it requiring construction input.

When they do move out of their domestic markets it is reported that many continue to use their tried and tested suppliers, often the same construction company that built their last domestic project. At the same time the need for local knowledge is recognised and the multinationals are quick to form joint ventures with local partners, an acknowledged trend in globalisation.

Comment

The multinationals are significant for a number of reasons: they are often first into difficult markets, there are lessons which can be drawn from their experience, they are useful as clients themselves since they are stable and they often promote worldwide standards which allow the non-local to enter new markets.

Who are the clients for the four sectors beyond the multi-national? In general they present no surprises: private sector and public sector, but in various guises. It is important to note that approaches vary on all aspects of the business and this will be studied in greater detail throughout the rest of the book. Increasingly overseas clients are demanding a total solution to their needs. In defence, for example, a country can request that a defence contractor supply the base that accompanies the patrol boats that it has ordered. The defence contractor is then expected to provide that service, often the majority of it construction-related, either in partnership with others or using sub-contractors. Their work then becomes part of construction's overseas output.

International markets can thus be classified and split in a variety of ways, but it is important that any classification aids clarification of the issues. Beyond the approaches of statisticians or the IMF the other popular approach is the geographical split into regional blocs such as Europe, North America or South America, or on a country-by-country basis. The disadvantage of many of these is their time dependency, which is more fully explained in Chapter 5. Business of any kind suffers from cycles of activity, and it is difficult continually to match supply with demand. There are therefore occasions when supply exceeds demand or vice versa, which has a strong effect on risks and rewards.

The key to a more generic analysis of the international market is concentration on the effect of risk versus reward. Understanding the risks and how they differ in different markets at different times is the key to providing a fuller understanding of international construction. Any other approach quickly becomes dated since it relies on constantly changing detail as opposed to a systematic approach possible through risk and reward analysis.

This systematic approach, discussed in detail in Chapter 3, can then provide a framework which applies to most international markets at most times. A framework of this type, however, needs frequent adjustment as it varies with time and location, but it is adjustment of detail and not the overall picture.

1.4   The set-up in South East Asia

It is important at this point to study a set of national markets to show the influence of the international on domestic markets. South East Asia is often viewed as the most international of regional global markets. It is a disparate group of nations which, nevertheless, are often lumped together as a single market. The first consideration is, of course, which nations make up this grouping. The main group of countries which form the Association of Southeast Asian Nations (ASEAN) provide the basis for one selection, although it is clear that there are omissions from the list in Table 1.5.

Table 1.5 South East Asia region (1993 figures).

Market	GDP (US$ bn)	Population (million)	GNP/capita (US$)	Construction (US$ bn)	Market (% of GDP)	Status
Brunei	4.0	0.3	13,450	0.4	10	Developed
Hong Kong	90.0	5.8	18,060	8.1	9	Developed
Indonesia	144.7	187.2	740	21.7	15	Developing
Malaysia	64.5	19.0	3,140	9.7	15	Developing
Philippines	54.1	64.8	850	7.0	13	Developing
Singapore	55.2	2.8	19,850	10.5	19	Developed
Taiwan	218.5	21.1	10,350	28.4	13	Developed
Thailand	124.9	58.1	2,110	22.5	18	Developing
Vietnam	12.8	71.3	170	1.9	15	Emerging

After Davis Langdon & Seah (1997) *Asia Pacific Construction Costs Handbook*. E & FN Spon.

Countries such as Myanmar, Laos and Cambodia are clear omissions. Less clear omissions are countries such as China, Japan, South Korea, Australia or other Pacific nations which increasingly trade within the region. Thus, the region has no clear boundaries and an artificial region is created, a sub-region of the Asian continent.

Table 1.5 shows a mixed bag of countries; one classified as emerging or low income, four classified as developing or middle income and four classified as developed or high income. Thus within this one region there are examples of each of the three sets of nations typically used in global breakdowns of national economies (OECD 2000). One from each of these will be used in the next section to further examine the state of the region in the mid 1990s and, more importantly, to examine the possible patterns which cause global analysts often to categorise this group of nations as one region.

Davis Langdon and Seah (1997), from which Table 1.5 is taken, make note of the poor state of much of the statistics for the region's national economies and there is therefore a need to treat the figures with caution. While this is true of many nations and regions it is unfortunately more so with developing and emerging nations.

The figures for GDP, population and the size of the construction markets themselves suggest more future potential than an existing gold-mine. By way of comparison, the 1993 figure for GDP in the UK stood at approximately £550 billion ($825 billion). At the same time, and in the same way, the figure for the construction sector was variously estimated at between £50 bn and £55 bn ($75–83 bn). Thus, the combined construction markets of the South East Asian region at $110 bn are only slightly bigger than a middle-sized European market such as the UK. However, as will be noted later, it is the rate of growth and the stage of development which contribute to the potential of the region (Bon & Crosthwaite 2000). As they are nations in the early stage of development, larger than average percentages of national wealth are spent on infrastructure and, by implication, construction. Thus, the combined GDP of the countries is actually less than the UK ($769 bn) but the construction market is bigger.

It is this factor which attracts outsiders to the market and, as a small regional market, the views of outsiders are important considerations in the perceived health of the construction sector of the region. Thus, despite the fact that none of the region's locally-based contractors made it into the world's top 200 contractors in the mid 1990s, the region still had an important part to play in international construction as a global competition zone.

1.4.1   The views of outsiders

The main global players in contracting and consulting in the region, and, to a lesser extent, the main building material producers can be identified from various sources. As sectors they exhibit very different traits across the region, which helps to explain the attitude and development of internationalism in the region.

The contractors

The big three exporters of construction influence, advanced techniques and management skills are generally taken to be the Americans, the Japanese and the Europeans (see the case studies in Chapter 6). The export statistics for these three show the importance of Asia to their global trade (Table 1.6).

The closest of the major nations, Japan, undoubtedly benefits most from the markets of South East Asia, whether through aid-

Table 1.6 Exports by contractors to Asia (1996).

Contractors	Value of orders to Asia (US$m)	Size of home market (US$bn)	% of Asian to home market
Japanese	16,910	550	3.1
American	5,494	710	0.8
European	9,461	790	1.2

Source: Thorn *et al.* 1997

supported work, following Japanese corporations into the region or in open competition. In fact Japanese contractors were reporting that 40% of their overseas turnover was sourced in the region by the late 1980s (OCAJI 1991). All three areas, however, had sizeable orders in the region. The British, for example, report that 15 of approximately 50 UK contractors active overseas had a major presence in the region (NCE 1996).

Comment

There is no doubt that in the 1980s and early 1990s South East Asia was viewed as a proving ground for any contractor with global pretensions. Although the traits of national markets varied, as will be illustrated in Case Studies 1.4 to 1.7 at the end of this section, the region was noted by all as being vibrant, with plenty of opportunity. Extremes existed from taking forward half-developed schemes in less competitive situations to full competition in some of the world's most competitive markets. This intense competition in Hong Kong probably led to its airport being procured at less than original estimates.

The consultants

The Americans, the Japanese and the Europeans (see the case studies in Chapter 6) all have a strong consultancy presence in the region. Again, the figures shown in Table 1.7 relate to Asia as a whole.

The closest of the major nations, Japan, has a poorer record possibly because of a less well developed consultancy industry and the importance of English as a language of contract. The USA and European results are sizeable given the distances from markets.

Table 1.7   Consultant services to Asia.

	Value of orders to Asia (US$m)
Japan	499
US	2,367
Europe	1,417

Source: Thorn *et al.* 1997

The British, for example, had a major presence in the region with 51 firms in Hong Kong, 42 in Indonesia and 33 in Singapore (NCE 1995).

Comment

Of the three sectors it is truly international consultants which appear to have benefited most from the demands, the opportunities and the openness of these markets to their products and services.

Building material production

In sharp contrast to the contracting and consulting sectors, global players appear to be less dominant in the region, chiefly because domestic players have fared well and brand names in this sector are less visible. Constructional steel is possibly the main exception to this, where the global big players have profiles and economies of scale at a global level. In the case study markets, for example, Malaysia reported one building material company and a number of conglomerates with sizeable building materials interest in its list of top 50 companies. In Singapore three building material specialists make it into the top 50 lists.

The economics of construction material sourcing, where transport costs are a factor, the abundance of natural resource and a degree of protectionism (see note in Chapter 4) probably all contributed to this picture in the mid 1990s. However, as noted in a case study in Chapter 6, the global construction material players are increasingly interested in the markets of South East Asia.

Comment

There are differences in the patterns across the sectors within South East Asia, patterns which in turn are affected by national variation in demand and supply. Some sectors, such as building material production, are more difficult to research than others as detailed in Chapter 6. The figures in Table 1.8 provide a clear indication of some of the national differences. Table 1.8 uses information as follows:

■ Davis Langdon and Seah (1997) for the public–private split, the 1995 cost factor (based on office costs) and the number of contractors
■ Franklin and Andrews (1996) for the second cost factor
■ Davis Langdon and Seah (1995) for the Vietnamese figure in the first and last columns.

Table 1.8 Construction–market characteristics in South East Asia.

	Public–Private	Cost factor (1995 & 1996)		Number of contractors
Hong Kong	40:60	94	123	3,700 registered
Singapore	59:41	104	108	3,890 registered
Malaysia	23:77	54	79	6,991 large firms
Vietnam	53:47	73	96	364 state + 1000 private

Singapore has twice as many contractors per head of population as Hong Kong. Vietnam had the unique characteristic of state run contractors. The cost factor figures (where both sources use the UK as a base cost of 100) show big variation across sources but it is important to note that Vietnam, the poorest, is viewed as more expensive than Malaysia.

The private–public sector split provides an interesting snapshot of variation. Free-wheeling Singapore has the highest percentage of public works, which as we shall see in Chapter 5 is a figure noticeably higher than most developed countries.

Case Study 1.4: Hong Kong

Viewed as a developed market Hong Kong has seen a huge transformation in recent years as illustrated in Case Study 1.1 on

Hong Kong Airport. Construction work loads have been kept high as a result and the competition regime is reportedly open, although problems occur, as a case study in Chapter 4 indicates. Interest from foreign contractors and consultants is intense, aided by the market's strong export–import culture. Local players have benefited from the growth and international links, with Gammon (now part of Skanska) the biggest regional player in South East Asia, Maunsell (a case study in Chapter 8) with a worldwide network run from Hong Kong, and Paul Y, a local contractor, which has now moved into the global top 200 (ENR 2000).

Case Study 1.5: Malaysia

The rapidly developing market of Malaysia has, like Hong Kong and Singapore, been viewed as an attractive investment location for much of the early 1990s. This caused a consequent boom in construction work with prestige projects such as Petronas Towers and early use of design build finance and operate contracts. The Malaysian government has placed many restrictions on foreign company activities (see note in Chapter 4) although this appeared not to reduce interest. Most major Japanese, Korean and European international contractors were represented and winning work in the country in 1996. The bigger Malaysian contractors meanwhile were beginning to venture into other countries such as Indonesia.

Case Study 1.6: Singapore

Often seen as competing with Hong Kong as the preferred hub of Asian operations, Singapore has likewise experienced a big boom in construction for much of the 1980s and into the early 1990s. The strong export–import culture has aided internationalisation and again most major Japanese, Korean and European international contractors were represented and winning work in the country in 1996. Its strong anti-corruption reputation has often assisted its attractiveness in the view of outsiders but competition has often been reported as very intense.

Case Study 1.7: Vietnam

Viewed as an emerging market Vietnam had been the subject of intense study (REMIT 1991; Davis Langdon & Seah 1995), but it continued to show many difficulties for foreign companies in moving into the country. The state run culture, its bureaucracy and the gaps in commercial law have hampered development, and advice for foreign construction firms has often been to work only on externally funded work. This in turn caused a reduction in the interest of outsiders during the 1990s despite the clear potential of the market.

Fig. 1.1

Fig. 1.2

The land area of Singapore has increased by 7% as a result of reclamation. Fig. 1.1 shows a completed project at Jurong Island (Phase 2) and Fig. 1.2 shows a newly completed project at Tuas view. (*Photos courtesy of Penta Ocean Construction Ltd, Tokyo, Japan.*)

Comment

The four nations are clearly different in development, culture and approach to internationalisation. There are clearly winners and losers in attracting investment within the region, although growth, an opening-up to global markets and other new developments brought pitfalls. This vibrant region of the early 1990s suffered a collapse in confidence in 1997/98 (see Case Study 5.1 in Chapter 5).

1.4.2 The future potential

After the trauma of the collapse in 1997/98, however, the region recovered and is once again offered as attractive. The benefit of hindsight is a wonderful gift and it is now clear that the region was overheated and overcompetitive during the early 1990s. A review of the region (*Economist* 1998) identified six problems thought to be common across much of the region:

- High investment, often in poorly planned projects, causing overheating
- Government intervention distorting markets
- Inflexible economies often through hidden government intervention
- Lack of transparency in corporate governance
- Lack of balance between long-term thinking and short-term results
- Shortages of skilled labour forces

Comment

Much of the above is speculative, i.e. there is no clear evidence that all these factors were directly damaging to the economies of South East Asia, although as discussed in Chapter 5, section 5.2, perception is almost as important as fact in international markets.

Thus a complex picture of a complex region arises, with more differences than similarities, and yet it still continues to be viewed as one linked market (*Economist* 1998).

1.5 The differences from domestic business

Having looked at the international–national twist it is also worth a brief look at the project end of the business and the differences likely to arise between domestic and international projects. Technical issues will not be studied in this book but there are a range of other issues which may vary.

A domestic construction project will require the following items for completion on site: plant, labour, materials, relationships, design to standards and willing client. An international project will require the same items. Viewed in this broad abstract manner there is little difference between domestic and international and this must not be forgotten. It is the detail which differs.

Case Study 1.8: East European roads

The former communist countries of Eastern Europe have spent much of the 1990s turning around their economies and preparing for integration into the European Union. The region's infrastructure is poor and road-building has become a priority, although funding is a problem since the governments have insufficient capital to fund it or credibility to guarantee payment for work completed. Thus progress has depended on drawing in aid support, Western banks and Western European construction companies together with private finance packages.

The European Bank for Reconstruction and Development (EBRD) is a prime mover in the work, acting as the agent for the EU (Mylius 1998). It reports that finance is the major problem since the region is poor and project returns of 10% to 15% demanded by the bank are not always feasible. Privately financed concessions completed to date in the region have featured many international firms such as Bechtel (USA), Bouyges (France), Strabag (Germany), NCC (Norway) and Impregilio (Italy). All report extensive use of local labour, plant and materials since they are cheaper than importing technology. Progress has been rapid on these projects – an indication of political support for the progress and no major technological problems that could not be handled.

Comment

The mix of development bank funding and international players but with significant local input is typical of what we would expect in international construction, with the location almost secondary to those main ingredients.

Looking at each factor separately, plant does not vary widely where the market is unprotected, i.e. the construction plant market is a global market in much the same way as other vehicular markets. The main exception to this is where protectionism occurs, which favours locally produced plant often made by inefficient makers. The rest prefer the most effective or efficient available and this has led to globalisation.

Manual labour, where people are used instead of machines, does not vary greatly in theory. In practice, there are differing rates of productivity and cost which can be exploited. This is more difficult than it sounds since economic theory suggests that differential rates of productivity and costs are quickly exposed in the open market and the difference quickly disappears. Good market intelligence and flexibility are required.

Design procedures, attitudes to time, standards and materials can vary, with any differences usually explained by differing engineering requirements or client views on quality. The other exception is in temporary works where local cheap materials, which can vary widely, are in use. A notable example is the bamboo scaffolding prevalent in much of South East Asia when compared to its equivalent in Europe, steel scaffold.

All of the above can be factored into a project preparation and analysed constructively. The great unknown is the relationships. Experience suggests that relationships are almost always more complex in international situations, the result of culture, tradition and ignorance. Other factors which strongly affect risk are the types of client, the contracts, the types of project and the availability of government assistance, all of which are discussed in more detail in following chapters.

Problem solving exercises

(1) Identify an international project from a recent journal or magazine. Identify the differences highlighted in the project description from your own experiences. Classify the differences under plant, labour, material, time, cost and quality/standards, sociocultural or climate.

(2) Review the internationalisation of your own country in the current year. How much has occurred? Why or how has it occurred? (See Appendix, Hints and Model Solutions for Problem Solving Exercises.)

2 Knowledge is Power

2.1 Introduction

Chapter 1 covered much of the basics, although in very simplified terms. It is now clear that international construction covers a broad spectrum of activities. Preparation for international activity involves a tension between:

(1) Having to gather as much information as is available to cover all of the options but then
(2) Focusing on only what is necessary for a particular project or operation

Information is vital in a complex business and it needs to be collated, structured and analysed. However, there is a balance between the need for focus and the need for broad information. A basic initial assumption must be that all relevant information available is necessary unless its value has been challenged. The information needs to be organised in a manner which will answer any why, what, how and where questions which are likely to emerge as a plan is developed to tackle the international business. Armed with sources for all of this information the follow-up work should then be selective, to focus on the most relevant information and, in a sense, to ignore other information where possible in any given situation.

A very well developed international company will have a clear strategy linked to clear markets or countries and it will have well defined sources of information for its needs. However, for many others, international opportunity will often be random in nature and time-scale. Any number of questions can arise: Should a business be built on the back of one opportunity? Why chase one project in, for example, Russia rather than Hungary? Who is likely to be the competition and what are the chances of winning the work? How much money can be made from a particular situation or opportunity? A well-researched business plan will provide many, but not all, of the answers.

2.2 Immediate sources of assistance

There are a number of sources of information which are useful, often free and tailored to the specific needs of construction. Government departments and trade associations devoted to construction are the most obvious initial source. However, before reviewing the actual information available it is worth reviewing the roles of governments and trade associations so that the value of the information and other assistance can be assessed.

2.2.1 Government

Government has a number of interests in international construction, with four loosely definable roles (Table 2.1).

Table 2.1 Government chief interests.

Interests	Form of support
(1) Economic	provision of general assistance
(2) Funding	provision of funding
(3) Statistics	provision of information
(4) Client	acting as client

The first and primary role is its interest in the success of the nation's economy, success of its exports and general diplomatic success. There are various functions that governments carry out in order to assist this success (see Case Study 2.1).

The second role, funding, comes from government's collection and retention of tax and other national funds. Civil servants often have some discretion in spending money where it will assist national interest, and construction interests can benefit from this 'largesse'.

The third role, statistics, is based on government's collation of the nation's vital statistics in order to assess the national progress. The information collected has a variety of functions that can be useful or make an impact on international construction.

For the fourth role, the emphasis shifts from information and assistance towards government fulfilling some form of a client role. Governments can be either a single body client themselves or they can join with other governments to provide a multigovernment approach. Much of this client role often comes in the form of aid packages, which are discussed in Chapter 9, section 9.3.

Not all governments have the resources or the desire to cover this full spread of services. Less developed nations, for example, are unlikely to provide direct funding for budding exporters and will instead concentrate on providing information to attract inward investors.

What is the value of all this possible assistance or, as some would claim, intrusion? The value lies in the fact that government wants to play a role in the first place and is thus usually willing to help construction companies when working abroad. But the actual material value depends very much on:

(1)   Timing – e.g. the business cycle
(2)   Location – e.g. support can be problematic in developed host nations
(3)   The specific part of government involved

There are typically a variety of departments within government which have a role to play in international construction, some better at assisting the private sector than others, and a lot depends on their agenda. We will use the UK government as an example, with a list of the various departments and an outline of their roles:

Case Study 2.1: UK government support structures

The Department of Trade and Industry (DTI) in conjunction with the *Foreign and Commonwealth Office* (FCO) through its new arm, *British Trade International,* looks at trade in general. It has a set of country desks to help export promotion to each individual country. The desks have expertise on their specific country, arrange visits and briefings and provide help points. Their advice is not intended to be specific to any one industrial sector although inevitably they do build up specific expertise.

The DTI is the lead government department for export promotion and it has an extensive website (DTI 2000) outlining the services available at the department. This includes directories of contacts, large collections of statistics, country profiles and lists of projects available from funding or aid agencies. The website also provides links to other Internet sites providing further information on specific opportunities or countries.

The Department of the Environment, Transport and the Regions (DETR) has a wide range of activities within its remit. Part of this is the 'sponsorship' of the construction sector. Sponsorship is a peculiar government phrase which covers legislation and regulation for the industry, support for the industry in general market improvement, research on specific aspects of the sector and through export promotion. Thus, there is specific construction export support at DETR and country specific support at DTI creating a matrix type effect of support.

The FCO, in addition to its role with British Trade International, has control of the embassies on the ground. They provide support to individuals or companies who are in a foreign country. This support is usually through the commercial section, although the full embassy can get involved. They provide information on the country, contacts and systems and often the embassies can provide accommodation for hospitality where required.

The Department for International Development (DFID) co-ordinates efforts on aid and the government's multilateral commitments. Aid comes in various forms; bilateral aid administered by one country for use in poorer nations, and multilateral aid administered through the European Union, development banks and aid agencies.

The Export Credit Guarantee Department (ECGD) which, in simple terms, provides export credit facilities and insurance for export companies against non-payment. The insurance is in theory limited to the risks which would not be covered by the private sector insurance industry (see Chapter 9). ECGD, in general, concentrates on major projects from all sectors, and this is probably true of many of the main central services.

Government in the UK has also set up a network of regional public body support organisations to help small and medium sized enterprises in their endeavours to grow and be successful. Often this support extends through to export promotion and support.

Comment

It is clear that the different departments fulfil a variety of roles. These roles are, in theory, brought together through the FCO in the target market and by the DTI in the UK (DTI 1993). A holistic approach is, however, seldom requested as most businesses seek to go it alone where possible, or use the government department as an initial source, preferring to collate commercially sensitive detailed information themselves.

There are equivalent departments in many other countries. For example, the ECGD equivalent in Japan is the Japanese Export-Import Bank (JEXIM), in the USA it is the Export Import Bank (EXIM), in France it is Coface, and in Germany it is Hermes, all of which are run along similar lines to ECGD, as we shall discuss in Chapter 9. It has been reported that many of these organisations lose money in their insurance activity, while ECGD manages to balance its books. While this may sound a small difference attributable to bad management, the net effect is that loss-making insurance equates to a hidden public subsidy (and it may be the result of good management!).

Case Study 2.2: CIA website

A good example of an information source which is available across the globe is the famous CIA (Central Intelligence Agency). The CIA, while more famous for its spying activities, provides an extensive amount of publicly available information through its on-line world factbook (CIA 2000). Detailed country profiles are provided together with updates from US embassies across the world. Project lists and economic guidance are also available

Comment

The power of the web is the subject of Case Study 2.4, Czech Republic, and websites such as this one provide a strong foundation. The CIA, despite its reputation for spying and detective work, has an extensive website with information which provides much of the basic background on international markets. It is clear that the US government through the CIA, its embassies and other government departments provides a first-class information service for its construction players working abroad.

2.2.2   Trade associations

While governments provide some of the background information, they would prefer industry itself to provide the support services. Trade associations provide ideal vehicles for this form of support. They are viewed as neutral, since in theory they are not biased towards one specific company, although problems can occur when there is one dominant player in a particular sector. They are also a useful platform for sharing information or for shared effort to persuade others on specific issues.

Each trade association is different although they share a number of common roles. One of their main tasks is to lobby government for the benefit of their sector within the industry. A second role, closely connected to the first, is the discussion of 'level playing fields' and the examination of the fairness of the market or regulatory environment, although at times it is suspected that some trade associations take this one step further and discuss how to actively skew the market in their favour.

A third role is in promotion of structures, processes and cultural change which will benefit all within their sector. This involves the gathering and dissemination of information and providing a base for discussion. This is a very useful function since, in export promotion, the costs of common market research particularly for smaller markets can be shared among members. The trade association's perceived neutrality aids this process.

A fourth role, which sees the greatest overlap with government services, is the gathering of statistics on behalf of their members. Trade association statistics are often viewed by the industry as being more objective and more up-to-date than government information, although that can be debatable and, besides, the sources are very often the same.

The fifth role is to actively promote their individual members. Again, in export promotion, this is often through trade missions focused on one particular country and sector of the industry. Just as with government departments, trade associations vary in style, content and usefulness. The best are active in the development of their roles and the promotion of their members, while the worst are mere talking shops. All of them involve, in one form or another, a group of companies getting together in a focused manner to look at a specific part of a specific industry. Within the UK construction industry, government recognises over 100 major trade associations representing some aspect of the industry. Within the export related sector there are a number of prominent players.

Case Study 2.3: UK trade associations

For the contractors there is the *Export Group for Constructional Industries* (EGCI 2000) which has a small closed membership. The members are in general the UK's biggest contractors and it sees its chief role as lobbying both UK national government to provide more support and the European Government to provide more level playing fields with regards to work practices, contracts and aid and trade provision.

The European Construction Institute is another contractor-based organisation. It is more proactive in the cultural change type of role and has extensive links with the offshore industry.

There are also two trade associations for consultants: BCB and ACE. The *British Consultants Bureau* (BCB 2000), a very proactive organisation, is renowned for its gathering and dissemination of information for and on behalf of its members. It is also an effective lobbying organisation with a particularly good record in Europe. The *Association of Consulting Engineers* (ACE 2000), by contrast, is a much wider association not focusing specifically on export and international promotion. Nevertheless it is strong in gathering statistics and in lobbying for its sector. There has been discussion of a merger of these two organisations although as yet this has not occurred.

The building material sector is covered by the *Construction Products Association* (CPA 2000) and by a host of other smaller, more focused associations. The CPA has tended to concentrate on gathering statistics and promoting members through trade missions and visits. Fragmentation is caused by the sector being focused around a number of core subjects (often material based).

Overseas trade associations

While the above information covers the UK it must be remembered that there are equivalent trade association services both across Europe (pan-European bodies and the European Commission) and in individual countries such as, for example, Japan. Table 2.2 gives some examples.

Table 2.2 Trade associations.

	Japan	UK	Europe
Contractors	OCAJI	EGCI	EIC
Consultants	JCCA	BCB/ACE	EFCA/FEACO
Building materials	(not known)	CPA	(not known)
Steel industry	KOZAI	BCSA	(not known)

OCAJI = Overseas Construction Association of Japan
JCCA = Association of Japanese Consulting Engineers
KOZAI = Japan Iron and Steel Exporters Association
BCSA = British Constructional Steel Association
EIC = European International Contractors
EFCA = European Federation of Engineering Consultancy Associations
FEACO = European Federation of Management Consulting Associations

Similar trends appear to emerge across all three areas in Table 2.2, the expanding role of consultants leading to ever wider groupings on one extreme and the diversity of building materials leading to smaller more focused associations on the other. All of these bodies deal in information and are often willing to provide information. One specific role for the pan-European bodies is the development of European standards for a single market (EFCA 2000).

2.3 Further sources of information

The title of this chapter is *Knowledge is Power*. To date we have concentrated on the sources of official state-approved information. However, there are other sources of information. The most obvious is paper references, typically through journals and magazines rather than books.

Case Study 2.4: Czech Republic

As an example of the paper references which quickly become available on a particular country a study of the Czech Republic in 1997 revealed the following:

■ A quarterly round-up of the Czech market from the UK Embassy (1997)

- A quarterly report from the US Embassy (US Department of Commerce 1997)
- A report on the Czech Republic from a building services research association (Holley 1995)
- A report on the Czech Republic from a quantity surveyor company (Barrow & Lawn 1996)
- General costings information (Davis Langdon and Everest 1995)
- The *World Competitiveness Yearbook* produced annually (IMD 1996)
- Various articles on the Czech Republic in the *Economist* and *Estates News* magazine (*Economist* 1995, 1997a, 1997b)
- Annual reports from European Investment Bank (EIB 1996) and the European Bank for Reconstruction and Development (EBRD 1996) which were available at the time on the Internet.

Comment

A huge amount of information is rapidly available on many markets, much of it accessible without a visit to the market. There is a mix of public and private, construction and non-construction, specialist and general all bringing useful information on the market. It is important, however, to make sure that the information is up-to-date and correct. Cross-referencing is particularly useful in this respect.

International construction, as we shall discover, is a complicated subject involving a wide range of factors. Many of these can change almost overnight so books on the subject can quickly become out of date. Thus, the journals and magazines which come out on a regular basis can provide more useful current material than books.

While construction is viewed as a technical subject some of the best sources of information can be financial or other more general journals. The growth of private finance initiatives has prompted financiers and lawyers to take more interest and this has created a source of articles on international construction subjects, often for the layman and often very readable.

Experience is another source of information but it is often underestimated. The experience of actually visiting a country is very useful. Trade missions with a group can result in a better view of the market since there is a collective gathering and analysis of the information available.

The funding bodies are also keen to educate potential suppliers of the needs of specific markets, projects or sectors since they perceive that a well-educated supplier is a better supplier. These bodies, which include host nations, other nations, multilateral agencies and financiers, all provide information, which is often free.

Finally, often the most important source of information is a local partner. They can provide critical information on the situation on the ground and the details of what is achievable, and they are particularly useful at the estimating stage.

Case Study 2.5: Web-based referencing

The Internet has become a very important tool. A list of useful sources will always quickly go out of date and so the example list below must be viewed in that context, although it is still a useful starting point:

www.worldbank.org
the website of the World Bank, includes excellent country profiles on most countries with explanation of infrastructure funding, and an explanation of spending on infrastructure
www.imf.org
the source of the World Economic Outlook and articles on background economics
www.oecd.org
global research on industry, which can be of interest to international construction players, although it is seldom aimed directly at the sector
www.ebrd.org
and other development bank sites provide lists of aid supported projects
www.brittrade.com
the website of the UK's Department of Trade and Industry export arm. This and its equivalent in other countries provides basic support information for exporters
www.cia.gov
the website of the Central Intelligence Agency, which gives a range of publicly available information – see Case Study 2.2

www.eurunion.org
the website of the EU with details of projects, legislation and reports from the European Investment Bank
www.worldconstruction.com
provides European press cuttings on the sector
www.ft.com
the website of the *Financial Times*, with occasional articles on construction
www.enr.com
the website of the *ENR* magazine
www.intlconstruction.com
the website of *International Construction* magazine
www.imd.ch
the website of the *World Competitiveness Yearbook*
www.worldeconomicforum.org
the website of a rival to the above
www.transparency.de
the website of Transparency International, which monitors corruption across the world (see the end of section 4.4 in Chapter 4)
www.fidicdirect.com
the website of the body made famous for standard contracts
www.eicontractors.de
the website of the European International Contractors Association

In addition to the above there are a large number of construction companies' home pages.

Comment

Increasingly the web is a useful source of information. Beyond the above sources, which are generally free, there are many other useful sources, including management consultants and interested third parties such as banks or legal companies.

2.4 Identifying the effect of risk

The objective at this initial stage of information gathering is to identify as many risks as possible. The second stage of refinement is discussed in Chapter 3, when the information is focused on a particular situation or project, looking for methods to alleviate the risks.

At this point, however, it is worth briefly questioning why domestic players look for international work when their domestic market is stable, well known to them and money can be made. It has already been stated that the classification of countries by risk is a standard approach (IMF 2000), although terms of reference vary. Reference has already been made in Chapter 1, Case Study 1.2, to countries being grouped as being developed, developing and emerging, together with countries in transition. Hidden within these classifications are patterns of risk and reward. Table 2.3 shows an overview of how the risks and rewards should work out in general for developed, developing and emerging markets.

Table 2.3   Risk and reward.

	Risk	Reward
Developed	Low	Low
Developing	Medium	Medium
Country in transition	High	Medium/High
Emerging	High	High

Thus, by implication, companies from developed nations are seeking riskier work when working in less developed markets. This is a gross simplification since different sectors of business have different risk profiles, and changes to classification or the risk and reward profile can vary rapidly with time. However, developed markets would include Japan, the USA and much of Western Europe, where risk and reward are viewed as low.

Generalised figures must be viewed with great caution, although anecdotal figures quoted for across the developed world suggest long term profit levels of 2% to 3% in construction.

Developing markets by contrast would probably be classified as having higher risks but also higher rewards, although they include the popular markets of South East Asia. Again it is problematic to put figures to this although an article about UK-based international contractors in the mid 1990s suggested that they sought 5% profit from international business from portfolios which included developed, developing and emerging (Cooper 1995).

Emerging markets have, in theory, the biggest risks and rewards. In addition to the technical and project risk that you would see in developed markets there is also the problem of political risk and much higher financial risk, i.e. a fear that payment will not occur. Emerging markets are likely to include, at the time of writing, Russia, Africa and parts of Latin America.

Having classified countries on this basis national governments then assign a risk factor to any trade between themselves and a target country. This forms the basis of trade and often aid policies. The most obvious manifestation of this is through state supported export credit and insurance policies on exports or major international projects (see Chapter 9).

This explanation of risk and reward is also the key to answering the earlier question of why strong domestic players look to expand internationally. It can often be when:

- Their market dries up or the domestic market becomes too small for the ambitions of the company or
- They wish to change their risk profile or spread their risk. Shareholders often demand greater returns than those available from traditional maturing markets, and international markets often offer that opportunity

The trick, of course, is to manipulate the opportunity so that a high risk, high reward situation can be changed to low risk but continued high reward. Comparison can be made with individual savers' approaches to savings portfolios where the risk-averse invest their money in low rate savings accounts but the risk-friendly go for shares which may fluctuate wildly. However, the opposite can also be true. It was long believed, for example, that the UK was a case of high risk but low reward as there were too many players in a shrinking market.

There are a number of publications available which try to calculate the risk for each individual country. This country risk is then added to the project risk in the assessment of individual projects (see Chapter 3).

Often access to the best information, government support and interaction with official approaches to risk assessment is available mainly to large companies who have both the time and the staff to devote to analysis of this type, where resource allocation is not a limiting factor. Many small companies do not have such luxury, and thus there is a fundamental difference in the small company approach where resource *versus* risk is often the limit rather than risk *versus* reward. Although risk will be studied in more detail in the following chapters, it is useful at this point to note the difference in how companies deal with the subject. Large organisations can afford to be slow and bureaucratic and are willing to take the measures which will see their work receive grants or guarantees from government, thereby reducing the risk. By contrast, smaller companies have neither the influence nor the resources, and yet there are successful small companies in international construction.

Small companies therefore need to establish other advantages by, for example, being quick and nimble since they will not be able to compete on resource. The risk needs to be managed in a different manner by limiting the resource going into an international project. There are many examples of this, with one case study in Chapter 8. There are many anecdotal stories of small, often specialist, companies competing internationally.

Thus, it is clear that there are no hard and fast rules for success. The approach taken has to fit the circumstances for each situation and company. In all circumstances however it is necessary to insure against the risk involved in the work or setting up the work. That insurance can take various forms for international construction (see Chapter 4):

- State insurance is available through ECGD
- Local staff and local partners are vital
- Contracts are often seen as the most important but rarely have any great value beyond outlining commitment in international jurisdiction
- The payment schemes which are often closely linked with the contracts.

Problem solving exercises

(1) Look at your own country now and assess (a) government support and (b) trade associations. Which departments or organisations are involved and what roles do they play?

(2) The equivalent of the CIA in the UK is a part of British Trade International within the DTI. Pick another country and find the equivalent. What information is available on the Internet?

(3) Choose one country and in thirty minutes on the Internet gather as much information of value as possible.

3

The Tools of the Trade

3.1 Introduction

It is clear from Chapters 1 and 2 that the international construction business involves many complex information requirements, and the quantity and quality of the information are important considerations. To facilitate discussion the wide-ranging characteristics of national markets are often simplified. The importance of standardising the approach and the information stems from a need to compare vastly different situations. Standardising the analysis framework is designed to maximise the ability to understand and analyse. At the same time, it must be recognised that each market and each set of information will be different. Companies are often faced with a situation where they are asked to look at one or two projects of a similar nature but in different countries or regions. A classic case of this would be an invitation to bid for the building of new McDonald's restaurants in, for example, Poland and Spain. Such a client would probably want a standard modular design adapted to local conditions. However, although the core building would be similar, the business, planning and regulation environment for contractor or designer would be very different. How can they be compared? Which one should a company bid to win if they have limited resources but a desire to best please the client? This again is a problem which is common across the whole of the business spectrum and not just specific to construction.

In this chapter attention will turn to a series of tools which are derived from standard MBA texts on toolkits for business analysis. There are many advantages to this; for example, it allows us to structure information, to use methods which provide easy presentation, and increasingly it is helpful to use 'MBAspeak', a standardised and powerful language across the globe, with many people aspiring to speak the new tongue.

There is nothing very magical about an MBA toolkit; in fact the main criticism is that it often represents an oversimplification of the facts. Most of the tools, however, can be used at differing levels of complexity as accompanying references explain, although the complex methods have their critics.

Case Study 3.1: Business analysis as a profession

A six year survey of senior management at large corporations revealed the extent of disillusionment among practitioners towards management consultants and their analytical tools, many of which are derived from MBA schools (*Economist* 2000). Companies reveal that they are confident about only the most simple of techniques. Jargon, complexity and queries about actual results have caused them to stop using many of the most fashionable techniques in recent years.

Other surveys regularly report high failure rates of other favoured implementation techniques of analysts and consultants. Mergers and acquisitions, for example, have a 60+% failure rate (*Economist* 1997).

Comment

Analysis and implementation are the two steps to successful business planning after information gathering. Remember that complexity does not always bring increased accuracy, and it is important to assess the credibility of any tool before using it. The simple objective at analysis stage is to ensure that essential information is structured in a presentable, easy manner and to avoid missing important bits of information so that clear implementation strategies can be formed. The analysis tools identified in this chapter are among the most widely used. They are presented in simple formats, without the more complex quantification methodologies which can be used. There is, however, no doubt that many would query their value while others would have their own favourites from the hundreds of tools available.

3.2   The generic tools

Analysis of the company and its environment involves analysis of a set of layers before bringing all the layers together. The external environment, the market and the inner workings of the company represent the layers. Here a tool for each layer is presented starting with the PEST (political, economic, social, technological) analysis looking at the big picture in its widest sense.

3.2.1 PEST

All companies operate within a global context. Much of what happens in the world has little or no effect on each individual company and its business. However, there are many factors or drivers which influence the company, frequently in a dynamic, complex, indirect or long-term manner. The most external of these factors are those over which the company itself has little or no impact. It can only react, either proactively before events change or reactively after the event. Acting before the event adds to uncertainty but it is believed that this provides companies with greater control of their actions.

Important factors are often likely to affect the wider industry rather than just one company, and they provide pointers towards possible opportunities and threats for the company. The widest environment of influence is likely to include the political situation in each individual country in which the company operates, the socio-economic conditions which create and shape its staff and clients' behaviour and the speed and direction of technological change which influences work practices.

The PEST framework (Shaw 1996) provides a mode of classifying and structuring information required in terms of political, economic, social and technological factors (or, as a variation on the theme, PESTLE, to include legal and environmental). It is basically a framework for structuring information as Fig. 3.1 shows, acknowledging that each of these factors represents forces which will influence any decisions arising.

The first step in developing the framework is the identification and classification of available information. This information can be collated through brain-storming, reference to key players or research of the market. A useful method is to construct an influence matrix (Shaw 1996). The aim is to collect information which points at factors which could have a significant effect on the business conditions in the market, short-term or long-term, although prediction is never easy.

The objective is to focus on a set of key factors and determine their possible impact. This is not a scientific exercise and it is

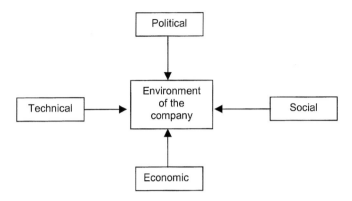

Fig. 3.1 The external environment. (Reprinted from Shaw (1996) and taken from *Foundations of Management: An Introduction to Strategic Management*. (Ed. J. Stiles), Henley Management College.)

important that a range of views are sought. It is also important that only key factors are assessed, since it is crucial that the exercise is manageable in scale. Although the key factors driving a national economy are not usually directly attributable to the construction sector it is often found that such factors will have significant effects on the sector either long-term or indirectly. Construction operates in a linked world despite our desire to work in isolation.

The main objective in the development of this information is to assess the impact of the factors and the priorities that this causes, i.e. practical action backed up with evidence has to evolve from the analysis.

To move from a random collection of thoughts on drivers through to a considered set of priorities requires a structured approach. There are numerous methods of presenting and analysing the data through a PEST framework. One method, shown in Fig. 3.2, concentrates initially on the visible issues, identifies the causes behind them and makes an assessment of their potential impact. The action required is assessed before establishing a set of priorities for a strategy.

It needs to be stressed that this type of analysis is not fail-safe; the world is too complicated and moves too quickly to be able to predict what will happen in the future. However, the exercise is

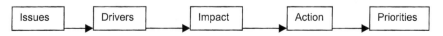

Fig. 3.2 Identifying the priorities.

worthwhile because it allows us to acknowledge how the external world influences the company, it promotes some strategic thought on where the company is going and it aids the process of joint decision-making, since consultation is crucial in the process.

Quantification of the impact is possible but since the exercise is neither in-depth nor scientific there is little point in pretending accurate quantification, and like all of the tools, the quantification is subjective in nature.

Case Study 3.2: Tarmac analysis (from PEST to action plan)

The worked example for most of this chapter is some analysis conducted in 1997 on a major UK company: Tarmac PLC (Mawhinney 1997). On the face of it, it seems a strange choice to look at a company which relied so heavily on its domestic market. However, there are four clear reasons for the choice:

(1) The company had a sizeable 25% of its business in overseas operations
(2) The author's past experience allowed a deep insight into this company
(3) More importantly, the analysis shows a company in the throes of change attempting to react positively to the pace and scale of external and international influences
(4) The benefit of hindsight has added an interesting dimension to this analysis

Despite the second factor above, the information used was all available in the public domain, i.e. no internal or confidential documents were used in the analysis. Thus, an analysis such as this should be feasible on any large company.

Tarmac was a leading company in the UK construction sector for much of the 1980s and 1990s. The analysis was conducted in 1997 when the business had a turnover of £2.5 bn ($4.0 bn). At the time there were five main strands of business: quarrying, building products, contracting, consulting and a US-based business. A sixth business, housing, was divested in 1996. The company was the largest of eight major companies spanning the UK construction sector and played a major role in contracting, consulting and building materials. It had a reputation as a traditional, reliable volume player rather than an innovative company. The annual reports at the time, however, were signalling change and consolidation.

It had already gone through considerable change in the early 1990s. This reflected the difficulties of the domestic UK market which was viewed as low profit and suffering from the continued effects of a recession in the early 1990s.

A simple PEST framework (see Table 3.1) was developed for the company's five (six) divisions, using the expert opinion culled from a list of references devoted to the company's future. This basically showed the main forces impacting on the company, which at that time was dominated by its domestic business.

The factors identified clearly had different effects on the various divisions within the company; some positive, some negative. An extremely simple quantification method was used to illustrate the possible effects, by categorizing effects as positive, negative or neutral. This allowed a judgement on which sector was a priority, and which factors affected which part of the business.

Comment

The PEST analysis indicated in Table 3.1 is very limited but the main message is that it is dominated by domestic consideration. The one obvious area of international work, the USA operations, was used as a hedge against problems in the UK rather than as a platform for more proactive operations. Other international operations appear not to be considered in much of the high-level strategy development.

Perceived belief outside the company was that it was ignoring the reality of globalisation which would one day catch up with it (Barrie & Billingham 1996). In fact, at the time, the largest shareholder was a Swiss Bank indicating just how much international events were already catching up with the company.

What the PEST tool does is to allow us to study the wider external environmental influences, which will frequently be indirect and medium- to long-term rather than direct effects. For the next analysis the emphasis continues on the external environment but this time on forces closer to the company. These are likely to directly affect the organisation, often on a shorter time-scale.

Table 3.1 PEST Analysis for Tarmac (Mawhinney 1997).

External variables	Quarry products	Building products	Tarmac America	Construction	Professional services	(Housing)
UK Govt Policy on						
Environment	– + ecogroup target	+ + added services	+ 0 diversified	– + ecogroup target	+ + added services	0 0 limited effect
Partnering	industry plus	industry plus	no effect	industry plus	industry plus	limited effect
Economic						
Reduction in public spending	– + smaller markets	– + smaller markets	+ 0 diversified	– + smaller markets	– + smaller market	0 + private sector
Buoyant economy	positive	positive	no effect	industry plus	business plus	positive
Social						
Aging staff	0 0 limited effect	+ 0 flexible product	0 0 unknown	– + labour intensive	– 0 increased costs	0 + + limited effect
Aging population	neutral	neutral	unknown	new markets	new markets	new markets
Technology						
IT	+ 0 added efficiency	+ 0 added efficiency	+ + added efficiency	+ 0 added efficiency	+ + + added efficiency	+ + added efficiency
New specialisms	limited scope?	limited scope?	new ideas	limited scope for change?	led by IT change	led by IT
Overall result	+	4+	4+	+	4+	5+

+ positive effect
– negative effect
0 neutral effect

3.2.2 Porter's Five Forces Model

The state of the sector itself is another important factor in the company's environment. The Five Forces Model was developed (Porter 1980) as a summary of the main forces which determine how attractive an industry will be. This is a well-known model (shown in Fig. 3.3), frequently used by management consultants when analysing the state of an industry and its main companies. Each force has a number of components and can have a positive or negative effect on the industry. The combination of the forces and the balances between them, allows us to summate how great the competitive forces are likely to be.

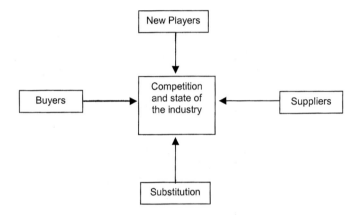

Fig. 3.3 Porter's Five Forces Model. (Reprinted from Michael E. Porter (1980) *Competitive Strategy* with the permission of The Free Press, a division of Simon and Schuster Inc. © 1985, 1998.

The central force, the *intensity of rivalry or competition* within the immediate market, provides the platform for gauging the intensity of competition. Porter's research has shown that there are a number of tell-tale signs which can be analysed to build up a picture from which it can be deduced if the industry is a highly competitive low profit market or if competition is low intensity and profits are fat. The tell-tale signs include:

Industry growth – there is an implicit assumption that growing businesses are the only type of healthy business. All the companies in one sector can grow only if the market itself grows. Thus, a growing market is viewed as a sign of good prospects for the companies in the sector.

Balance of competitors – many markets or sectors will have two or three main rivals and a number of smaller players in the market. More intense markets will have larger numbers of equally sized players, seen as an indication of an unprofitable sector.

Exit barriers – the barriers to entering and leaving the market are significant factors in the health of a sector. Industries which have high levels of investment or specialist knowledge bases make it difficult for new companies to establish a presence, or for existing companies to leave during bad times.

Expected retaliation – the strength of a response to, for example, price cuts by one competitor is another important parameter. A slow or negligible response as a norm may indicate an industry where all players have a comfort margin or, alternatively, may show complacency.

The *bargaining power of buyers* gauges whether the buyers dominate the market and dictate the pattern and prices within the market. Likewise, the *bargaining power of suppliers* can be assessed in a similar manner. Likely tell-tale signs include:

Volume of purchases – buyers or suppliers to the main players who control large volumes can have a very influential role since the flow of material or service provides them with a key lever.

Market share – buyers or suppliers to the main players who control a large percentage of the market can have a very influential role since they have a dominant position which can translate into a control of pricing.

Switching costs – low costs in switching from one supplier to another can be influential in reducing the power of suppliers.

Profitability – strong buyers or suppliers can, to a certain extent, impose a level of profitability on the main players.

The *threat of substitution* implies that there may be substitute products which could replace the traditional products in this particular market. The threat of substitution may also imply that there may be *new players* available to join or about to join the market and increase the competition, and again an assessment is necessary. Again likely tell-tale signs include:

Economies of scale – industries where there is a strong economy of scale factor (i.e. the size and location of plants, offices or suppliers is important) tend to have a limited number of players who can play the game. This can preclude new players.

Existence of close substitutes – products in particular have developed through the years as one product has replaced another, sometimes in a revolutionary manner. Mechanical watches were replaced by digital watches and the Swiss watch industry almost collapsed. Thus, the potential for replacement products can decide how attractive an industry remains.

Switching costs – the actual cost for buyers of switching from a mechanical watch to a digital watch is a factor in the speed of turnover from one state to the other.

Access to distribution – contractual binding, location or other factors which preclude competitors from sharing suppliers can be a factor in reducing threats.

The importance of these factors is in the degree of control a company can lever in determining its stance and price within this particular market or for a particular project. The bottom line is in establishing how attractive a market or a particular situation will be and whether it is worth competing or ignoring it.

The difficulty in assessment is that there is no right or wrong. All markets are different and different combinations of conditions can determine whether one or a small number of key players control the market. Thus, the analysis is subjective, and there is a degree of expertise in judging whether a particular market is attractive and competition is not too fierce.

Case Study 3.2 (cont.): Tarmac analysis

Again, we return to the analysis of Tarmac to explain the various factors and to explain possible quantification. It is possible to quantify the separate factors partially in a crude manner similar to the PEST analysis techniques. They can be quantified as high, medium or low forces for each (Table 3.2).

The analysis of Tarmac depends once again on its position domestically within the UK. In other developed countries there are typically 3 to 5 major players. At the time of the analysis there were 8 equally balanced rivals and zero growth in the UK leading, of course, to high intensity of rivalry. This would suggest an

Table 3.2   Forces in evidence in the UK construction industry in 1997 (after Mawhinney 1997).

	New Players	Buyers	Suppliers	Substitution	Competition
Characteristics	EU companies Utilities	Uncoordinated	Uncoordinated	None visible	Zero growth 8 similar rivals
Forces	Medium (high long-term)	Low-medium	Low	Low	High – too much competition

unattractive, declining market, much as city analysts suggested at the time. This leads to low margins and a view that the industry is not an attractive one, although it may also put off potential competitors or new entrants from entering the market.

An interesting factor is the view that buyer and supplier power was low. Fragmentation is a key factor in this result although the net effect appears to be the two forces cancelling each other out. The international threat from new competitors was a significant factor discussed in general industry terms but seldom mentioned in the context of analysis of Tarmac itself. The benefit of hindsight suggests this was short-sighted as is revealed later in section 4.2.

Comment

More sophisticated methods of quantifying the effects through Porter's model are possible. However, subjectivity limits accuracy and this author is not a great believer in pursuing apparent accuracy when the initial data can be open to interpretation. The benefit of the method lies in its simple approach to showing the pressures on a company within a market.

3.2.3   PARTS checklist

The PARTS checklist (Brandenburger & Nalebuff 1996) provides another framework for structuring information and checking that all vital information is included. It has, to date, provided less scope for quantification methods. Its theme is excellent since it seeks to identify competitors as both competition and potential co-operating bodies, trying to redefine all aspects of the market or an opportunity as having both positive and negative potential.

The framework revolves around the five terms whose initial letters form the acronym PARTS – players, added values, rules, tactics and scope. The overall intention is to think of business as a game, and to improve the game for everyone involved by asking the right questions.

Players – Which players do you work with? Can you co-operate and compete with them all? Are there different players to involve? Who gains and loses when you enter a market? Who gains and loses if the rules change? Who are the players and can they be made to play together rather than against each other?

Added values – What is the firm's competitive edge? Can it be increased? What is the added value of other players? Can the supply chain add more value than the company itself for a particular situation?

Rules – What are the ground rules for the market and, rather than accepting them, can they be changed? What rules help and which ones hinder? Who changes the rules?

Tactics – What are the tactics for competing? Is the game transparent or opaque?

Scope – What is the current scope of work accepted by the main players and, rather than accepting it, can it be changed? Is the game linked to others?

All of these are very useful questions, although they generally get lost in the heat of competition. The checklist is, in my opinion, useful for reviewing a situation but not a 'first order' framework such as PEST or Porter's Five Forces.

3.3   Internal strengths

Having reviewed the main external factors, work is required on reviewing the resource available within the company to devote to the project or business plans. It is often believed that a company's strengths are blindingly obvious (ask any gruff managing directors of a construction company), and a mechanical, off-the-shelf list is often produced to summarise asset or past experience. Frequently there is neither structure nor logic nor relevance to the current situation attached to this.

The author's experience is that this is often a poorly performed part of the preparation for action. There are too many 'sacred cows'; a company builds a prestigious bridge in 1985 and it still appears on its prequalification proposals in 1997 despite the technology having moved on, the personnel involved no longer being in employment and no further bridge work since that time. Why does it appear?

Possibly because the managing director was project manager at the time and nobody wants to offend him!

Again, however, construction is not alone in having this problem. It is common across many industries, and management consultants have made fortunes on the back of it and companies' inabilities to deal with 'sacred cows'.

Critical to good analysis is the acceptance that a company's strengths and weaknesses revolve around three factors: its financial or asset base, its people and its direction. A strong company needs strength in all three directions to progress. Too often analysis revolves around a quick check of the financial position, which is actually the most problematic of the three axes.

There are a number of tools which can help in this analysis, although the best known are tools which involve separate analysis of the three factors. This is an area well covered in management texts with a variety of schools of thought (for example Kay 1995; Kotler 1997) and will be further examined later in this chapter and in Chapters 4 and 5. At this point the theory will be skirted in favour of explanation through the case study.

Case Study 3.2 (cont.): Tarmac analysis

The analysis of Tarmac in this respect showed continued emphasis on domestic considerations. At this level the company was performing adequately. Although it had a worsening financial position, the decline was less pronounced than many of its competitors in an obviously poor market. The sales per employee, a measure of efficiency, were amongst the best in the industry suggesting a strong human resource base. However, a major problem was the mixed messages that were being delivered in terms of direction, with senior management advocating quality and innovation, and divisional management stressing volume and market share. Many MBA enthusiasts would see these messages as being incompatible. City analysts, a key external body, read the results as mediocre and thus put great pressure on the company's management radically to alter its direction.

3.4   Matching the internal and external

Having identified internal and external factors, the two sets of information need to be combined. How can we make the most of what is available internally to capitalise on the opportunities available externally? A SWOT analysis is a popular method. This provides an important contribution in developing a business plan to link and explain the implications by using the information gathered through the previous exercises to summarise the main strengths, weaknesses, opportunities and threats. The strengths and weaknesses are, in many senses, internal, while the opportunities and threats chiefly arise through the external analysis. It is very important to recognise that many factors represent both opportunity and threat or both strength and weakness (e.g. a young, dynamic management team is also an inexperienced one). It is necessary therefore not to paint a picture with obvious biases which can quickly cause problems.

The development of the SWOT table must be viewed as a presentational tool, the most important point being the analysis which accompanies the production of the table. The conclusions must include an explanation of the best match of strengths and weaknesses to the perceived opportunities, i.e. what is the best opportunity which can be developed? If one opportunity is already in mind, i.e. one project, the question turns to whether that project fits the needs of the organisation and does it represent a good opportunity?

Case Study 3.2 (cont.): Tarmac analysis

The SWOT analysis of Tarmac (Fig. 3.4) again hinges on domestic influences and shows the priorities arising from the previous PEST, Porter and internal analyses. It appeared that all opinion agreed on the threats to the company but not on the opportunities. A boardroom struggle at the time revealed two conflicting strategies, both with a desire to move out of traditional markets;

Strengths	Weaknesses
Overall market leadership	Poor margins and pressure from analysts
Dominance in aggregates	Recession hit construction division
Efficiency	

Opportunities	Threats
Fragmented across divisions	No strong recovery in UK construction
EU market opening	EU competitors
Growth of private finance initiative (PFI)	Finance required in PFI projects

Fig. 3.4   A SWOT analysis for Tarmac in 1977.

one was based on speeding up change and greater overseas operations and the other, eventually accepted one, was based on movement into related sectors such as facilities management and privately financed operations (Barrie & Billingham 1996).

Comment

The situation where two conflicting strategies emerge from the same analysis is not unusual and highlights the difficulties of drawing conclusions and predicting the future. The eventual decision not to concentrate on international influence initially appeared sound and after further restructuring the company was split into two: a construction services company and a quarry products company.

However, very quickly after the de-merger the stronger quarry products company was snapped up by a South African company, stripped of its senior management and thus became a minor part of an international conglomerate. The construction services company remained independent despite signals that senior management wanted it to be bought out. The strategy of moving into services related to construction was, in a sense, a declaration to change the scope of the company's activities (remember PARTS). Since strong balance sheets are critical to some of these activities it should also reduce competition because of the barriers to entry (remember Porter's model).

3.5   Setting the business direction

The summation of the various analyses completed to date plus added financial detail and some creative study of possible future scenarios provide the main parts of a business plan. Any conclusions of the business plan should be clear and unambiguous, i.e. yes, we will proceed with the proposal or no, we reject the proposal or specific actions must be taken before a decision can be taken. Good conclusions will be based on creative thinking and the analysis, addressing the opportunities and making best use of all the factors available.

The sustainability of the business is both short-term and long-term. In the short-term sufficient cash must flow through the business to pay the bills and pay for daily operations. Long-term there must be profit sustaining growth or future development of opportunity. It is important to recognise that the two of these can be seen to conflict. However, good management would see the need for both short- and long-term development in an approach that covered the whole of the business – the subject of management texts and guides.

An important element of this is the need for some sort of benchmark against which a measure of progress or success can be determined. A useful starting point is the initial business plan, since it should provide an idea of the aspirations and milestones that were initially envisaged. Most management gurus would advise against the idea of sticking rigidly to a business plan. It is important to have a flexible approach to business so that the company can respond to opportunity as and when it arises. Situations change and, with this, come varying costs, changing opportunities and new threats. However, even the most flexible management gurus (and there are plenty of them) would all agree that it is important to monitor the finances, and the financial element of a business plan is a useful starting point, even if it must be treated with caution.

Table 3.3 shows three content lists for good practice business plans, two from well respected management consultancies and one from business school notes (Atkinson *et al.* 1986; Henley Management College 1995; Ernst and Young 1994).

There are many other references on this subject, all with some variation in how they are structured, their emphases and the associated priorities. In general, however, the content is similar; the common themes are the need for a structured approach to the

Table 3.3 Business planning.

Atkinson *et al.*	Henley	Ernst & Young
(1) Targets and objectives	Executive summary	Executive summary
(2) Product plan	Organisation context	Background
(3) Pricing plan	Situation assessment	Products
(4) Market information plan	SWOT analysis	Market analysis
(5) Sales plan	Market opportunities	Marketing/selling
(6) Promotional plan	Assumptions	Manage/Organisation
(7) Structure/staffing	Objectives, strategy and action plan	Fund/finance projection
(8) Budget	Financial plan	Risk assessment
(9)	Contingency plan	Action plan

information, consideration of resource and finance, a set of trackable milestones and analysis of the competitive advantage.

Two of the three lists in Table 3.3 include risk or contingency assessment, the subject of section 3.6. There is also common agreement that too much information is as bad as too little information, and therefore structure and succinctness are critical. Many of the basics have been covered in this chapter and competitive advantage will be studied in further detail later in Chapter 5, section 3. In this section two areas will be further developed – creating a set of wider objectives with a set of measurable milestones and consideration of the finances – since these are key in sustaining the business after initial set-up.

3.5.1 Developing a set of wider objectives and a set of trackable milestones

The objective in developing a business plan initially is threefold:

(1) As a planning tool
(2) To persuade others of the merits of the decision to proceed
(3) As a means of monitoring progress and highlighting change required.

All three of these objectives are important and need consideration. It is relatively straightforward to set SMART objectives (suitable, measurable, achievable, realistic, timely) and therefore suit short-term needs and longer-term monitoring requirements. The objectives need to translate into a set of measurable milestones that can be used at operational level. The following case study illustrates how one company, although unrelated to construction, set up the links.

Case Study 3.3: GEC's famous matrix for planning (Kotler 1997)

General Electric is an enormous US based company which in the 1980s employed over 400 000 employees across 43 diverse divisions. Senior management had a very difficult time assessing performance and planning because of the diversity. They eventually produced a relatively simple matrix based on two concepts – industry attractiveness and business unit position.

By measuring all the businesses against a common set of variables under these two categories it was possible to set measurable targets and goals across the firm. Under industry attractiveness was a combination of market size, industry profitability, cyclicity, inflation recovery and importance of international markets. Under business unit position was market position, measures of competitiveness and relative profitability compared to competitors at the local level.

Comment

A matrix with the two representing axes allowed senior management to assess where all the divisions lay within their own particular markets, while still allowing an overview for the company as a whole. The combination assessed covered a wide range of business factors. However, models like this have been criticised for being over-simplistic.

Both changes in market and finances can quickly render the original business plan obsolete and this requires decisions on further planning. The company culture could demand a new business plan every time there is a change to the financial position or objectives need to be reset. Alternatively, culture or time constraints may impose a fire-fighting approach. Both serve their purpose.

Case Study 3.4: Anecdotal view of the British by the Japanese

An anecdotal view often heard by the author in Japan portrays British managers as having weak planning and consensus-building abilities, both of which are critical in Japan and often widely praised in management texts. However, the Japanese view is two-sided; weakness in these has meant that fire-fighting ability becomes crucial, and many British managers have been employed in Japanese firms for their believed strength in this competence.

Comment

It is dangerous to generalise in such a fashion, although experience in Japan leads me to understand the thinking. Obviously the happy medium is somewhere in-between, with sufficient flexibility to react quickly when necessary or to plan when replanning is required. That happy medium will accommodate reference back to the original business plan to spot tell-tale signs of significant variation from the original objectives or financial forecasts so that action can be taken quickly and decisively.

In both cases it is important that a culture of review, evaluation and monitoring the right factors be developed. In some of the references this is partly brought about through the action plans and through a heavy reliance (sometimes over-reliance) on financial planning.

3.5.2   Financial forecasting

This section will provide a brief overview of financial forecasting, a complicated subject which eventually must make reference to an enormous range of issues such as borrowing, taxation, etc.

Forecasting the growth of a business is a difficult exercise. There are many external factors which intrude on smooth growth and planning. The use of sales forecasts from the sales team or the adoption of a projection which is based on a rival's approach in similar circumstances are useful starting points. The sales or income generation must be grounded in evidence or expert sourcing, consistent across the plan, and the assumptions involved must be described in sufficient detail that they can be challenged or tested if need be.

Although central to the business plan it is important to remember that financial forecasts are only one part of it. There are a variety of forms of projection but they are all likely to contain the following elements:

Cash-flow – the basic calculation of money in, money out. The costs to the company of setting up the business, the on-going costs and the revenues are modelled in a spread-sheet. This can then serve three purposes:

(1) An indication of when the initial investment will be paid back can be calculated using the payback method
(2) The highs and lows of the cash-flow can be assessed so that bank loans can be arranged
(3) The average rate of return of the business can be calculated as an indication of the profitability of the business.

Discounted cash-flow – money has a value in time and this approach allows this to be taken into account. One dollar in your pocket 100 years ago was worth more than a dollar in your pocket today. The discounted cash-flow (DCF) translates future cash-flows using a discount rate into net present-day values (NPV). The rate is often the projected cost of borrowing the money but it may be adjusted to reflect risk, a desired profit or inflation.

Profit and loss forecast – involving the accountants in the wider picture. In the normal scheme of things the cash-flow would pro- vide a set of costs and a set of revenues. The difference between the two would represent the profit of the operation. In the real world, however, there are many things such as taxation, the value of assets and other items which distort this simple picture. Companies are obliged by law to provide a profit and loss account, and planning, whilst difficult, can be useful in international settings where tax regimes for example vary enormously.

As stated earlier, forecasting the potential growth is difficult and there are many unknowns which may render the plan quickly obsolete. However, references (Ernst and Young 1994) suggest that transparency and the use of expert sources represent the best way forward. Plans must, critically, show evidence of supply and demand. The capacity of the company must be referenced and must be consistent with the projections, competitors' capacity needs to be referenced since this is an important external pressure and the demand of the market needs to be referenced. In working with students in the past few years I have found a common fault has been the development of projections which are based on flights of fancy beyond the capability of the market itself.

By incorporating measurable milestones, financial targets and a set of clear goals the business plan becomes a very useful document because, apart from its value in setting the scene and completing analysis from which comfort can be derived, it can also be used as a benchmark from which future work and progress can be monitored. Delivery and action plans are critical in this respect.

Finally, what must be included in every business plan is a reality check. The analysis must have a section which outlines how the analysis leads to a situation that is workable, real and connected to on-going circumstances and demand.

3.6 Risk and reward assessment

The previous sections looked at the corporate background and the assessment of wider external and internal factors to determine the types of constraint and opportunity. We now move on to some tools which help us to analyse the scale of individual opportunity on projects. Although the level of project opportunity may often be perceived as representing a secondary analysis beyond the corporate level analysis, it is actually equally important. A quick explanation is necessary.

At a corporate level, a construction company setting up a business overseas will typically set up an office first. The costs for this may be $0.5 m (see Case Study 10.3 in Chapter 10), which if everything goes badly represents a direct loss). At project level, a major project may, for example, be worth $50 m. Although the funding is not initially company money the potential for loss or the exposure to risk is much greater until adequate steps can be taken to safeguard the company through contract or insurance arrangements.

The analysis of individual opportunity concentrates on risk and reward assessment. This is the analysis of the likely or possible benefits versus the potential for problems. The methods are global and, once again, they apply in all industries. Although they generally form a two-part analysis there are organisations which have attempted to roll them into one.

Reward analysis is often measured solely in financial terms and is a subject extensively covered in many good texts. We have already looked briefly at the basics of financial forecasting in section 3.5.2, but it is worth noting that the end result of the reward analysis will be a single value or range of values depicting the profit or cash arising as a result of the pursuit of the opportunity.

The second part of the assessment is the risk analysis. As with all the tools described in this chapter there are numerous approaches, many of them scientific in appearance. The method chosen here is, however, to stick to simple basics since risk analysis is about predicting the unpredictable and accuracy is impossible. An over-sophisticated model leaves the impression that accuracy is being achieved when that is not actually the case, and many fall into this trap.

There are a number of useful texts on this subject. One of the best was developed in the UK by CIRIA (Construction Industry Research and Information Association) (CIRIA 1994), who based their methodology on tools developed in the pursuit of improvement of health and safety in the industrial sector. This has produced a useful generic framework which can be modified. The CIRIA type of model has a number of simple initial steps: list all the risks, assess their consequence, assess their likelihood and look for methods of alleviation. Listing all the risks can be a very difficult exercise, e.g. it has been stated that PFI (private finance initiative) legal experts have identified 400 separate risks in larger PFI projects, all of which can require separate assessment and contracts to specify who is responsible. As lawyers, they would say that wouldn't they! However, in general, the combination of experience and a standard checklist can greatly reduce the list to more manageable proportions.

A review of some available checklists shows two main types: client and service-provider. The client lists of risks are generally wider in scope since construction forms only one portion of the risks to the client. Lawyers (Freshfields 1996) point to revenue risk, operational risk and construction risk. National Power (1997), a substantial client, add financial risk, legal risk and political/cultural risks. Freshfields suggest the construction risks can be split into two. The first includes ground risk, location risk, availability of utilities risk, access risk, legal risk and environmental risk, all of which lie with the contractor and designer. The second includes information deficiency risk, disruption risks, payment risk, planning and regulation risk, and it is suggested that these are open to negotiation between client and contractor particularly in PFI type projects.

For service-providers, i.e. contractors and consultants, the CIRIA checklist compares well with one produced by a major company which has now been superseded (see Table 3.4).

Table 3.4 Risk lists.

Commercial	Conditions	Delay damages
	Performance damages	Bonding
	Fixed price	Ground risk
	Design responsibility	Defects liability period
	Integration/interface problems	
Financial	Cash flow	Margin
	Profit downside	Payment security
	Extent of penalties	
Company	Technology risk	
Exposure	Technology experience	
	Partner experience	
	Single source suppliers	
Location	Access	Political stability
	Language	Dispute recourse
	Foreign exchange	Weather
	Labour disputes	
Client	Company experience	Current exposure
	Imposed delays/constraints	

Comment

The checklist in Table 3.4 shows elements of PEST and Porter's thinking (see sections 3.2.1 and 3.2.2 earlier in this chapter) but with a heavy reliance on contract and finance. The next step is to score each of the factors and this is where the two systems diverge. CIRIA's case is described below; in the major company example weighted ranges of scores are pre-assigned and a score is derived from this (see the problem solving exercise 3 at the end of this chapter).

Having developed a list the next step is to look at possible impacts of each risk. A quantification and explanation of each risk factor is called upon, in terms of the likelihood of the risk occurring, the consequence of the risk and the alleviation possible. The likelihood of the risk is the probability of a problem occurring within the construction period. The consequence is the damage which is likely to occur. Each of the assessments is scored or weighted.

In engineering terms risk is then defined as likelihood × consequence, and the CIRIA approach would suggest that any possible alleviation of the problem is removed from the equation (i.e. if the problem can be fixed or reduced with little or no effort then the multiple is reduced).

The exercise of developing an explanation of why and how the risk will affect the project outcome is as important as the scoring, often helping in the decision on quantifying the risk. Experience is vital at this stage since subjectivity can be a significant factor. There are systems where a set of experts are asked to score each risk factor and an average is derived. Alternatively, computer models are used to decide statistically from previous projects. My own belief is that the unknowns are such that dressing a subjective decision up as a scientific one is a problem, and I therefore prefer the consultation of experts as the better option, although others would disagree.

The definition of risk as likelihood × consequence is at odds with the definition of risk in financial circles. In this case it is defined as the likelihood of an object to behave erratically and there are some very complex methods of analysing such behaviour. For many years this anomaly and the double definition has caused little or no problem. However, with the onset of PFI where both sectors have a large vested interest it is clear that there will be a future need to harmonise at least the vocabulary if not the way of thinking.

The method of quantification suggested by CIRIA uses a scoring system of 1 to 5 which is viewed as being helpful in seeking objectivity. The scoring is shown in Table 3.5.

The importance of this assessment is not in the absolute figure derived but in the analysis that is required in reaching the

Table 3.5 Quantifying risk.

Likelihood		Consequence	
Frequent	4	Catastrophic	4
Probable	3	Critical	3
Occasional	2	Serious	2
Remote	1	Marginal	1
Improbable	0	Negligible	0

Reproduced by kind permission of CIRIA Ltd (CIRIA 1994)

conclusions. It is very important that the total developed from an assessment of this sort is viewed as nothing other than a vehicle to assess the risk. It is comparative rather than absolute, although with experience a view can be taken of its comparative value.

Both risk and reward analyses are subject to wide degrees of error, and sensitivity analyses are recommended. This involves testing the analysis using a range of possible assumptions to determine how sensitive the results are to change.

It is clear, however, that the CIRIA type of risk analysis has many uses and the author has used the tool in, for example, the analysis of inward investment. Again, it is a globally acceptable tool with globally acceptable terminology.

The question then arises of how to join the risk assessment to the reward analysis. Some companies carry out the two separately and then informally use one to inform decisions on the other. Another possible method of showing this is to make the two represent two axes in the same graph (Mawhinney 1999). Other companies have devised complex systems of assigning a value of reward to each value within the risk assessment (e.g. National Power). This is likely to lead to a high risk high reward, low risk low reward outcome (see Fig. 3.5)

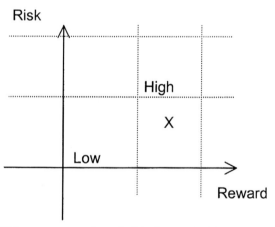

Fig. 3.5 Risk versus reward graph. X indicates the ideal – high reward, low risk.

Neither risk nor reward analysis can be classified as truly objective and the author therefore prefers the method of informally linking the two since this does not allow us to pretend that the absolute figures are scientific. However, there is scope for development of a simple method of linkage.

It is worth remembering that the tools introduced in this chapter
are simplistic, subjective, have many gaps and have been derived
from work outside the construction sector. However, given the
complexities of the business in international construction the use of
simple tools is more justifiable than any pretension that all variables will be modelled and accounted for.

Problem solving exercises

(1) A list of analysis tools has been provided in this chapter. Take
these or others of your choice and list all the disadvantages of
using them.

(2a) You work for a producer of heating and ventilation systems
based in Western Europe. The information below presents an aid
supported project opportunity which represents a useful opening
to the markets of Latvia but may not be sufficient in itself to
develop profitable business. Describe what steps you would take
to further research the market, source information and develop a
market plan. Describe the options for supplying a project like this
and any other follow-on work. (See Appendix, Hints and Model
Solutions for Problem Solving Exercises).

The project

	Population (million)	GDP/$ (bn)	GDP growth %	World Bank spending (million)
UK	56	$1,384	2	—
Estonia	1.5	$4.5	8	$125
Latvia	2.5	$6.5	5	$150

Latvian Airports, a state enterprise, has obtained funds from the
Global Development Bank for the reconstruction and development
of Riga Airport. The proposed project will require the following:

- Repair of the terminal building and provision of new addi-
  tional buildings
- Installation of a new power supply system
- Installation of a new communication system
- Installation of a new water supply and sewerage system
- Installation of a new heating and ventilation system

■ Modernisation of security, mechanical and emergency systems

The works are expected to take 24 months in total and the total funds allocated for design and construction are 28 million Deutsche Marks (DM) (assume 2DM = US$1). Prequalification and tendering will be to World Bank rules and guidance.

(2b) Prepare a financial plan.

(3) The following information has been provided on a project opportunity. Your company has a standard risk assessment model (see framework below) and it wishes to use this so that a preliminary decision can be made on whether to increase the resource provided for the project and proceed with more detailed preparation. Complete the assessment and report your conclusions (see Appendix, Hints and Model Solutions for Problem Solving Exercises).

The project

The project is a new 70 MW hydro-electric power station in Central Vietnam. The company has been asked to bid for the engineering, procurement and construction management of the mechanical and electrical engineering (M&E) works, with a contract value of approximately US$45 m. The project calls for the commissioning of two turbines, one to be completed in two years and the other following 3 to 6 months later. The owner has requested that a 100% financing proposal is required, i.e. the contractor will need to find a bank or export credit agency or finance it in some other manner during construction. The bid will be fixed price.

The form of contract will be the client's own conditions and the governing law will be the laws of Vietnam, although the contract language will be English and payment will be in US dollars. Advance payments of 10% are available for both equipment and construction management, although the follow-up payments are upon receipt of major equipment and on a monthly measure basis for construction management work.

The contract documents call for the following: $500 000 bid bond, performance security of 10% of contract price, recognition of the usual *force majeure* situations, a defects liability period of 12 months from handover although it will be specifically 6000 hours' use for the actual turbines. There are no retention clauses specified although liquidated damages for delay can be up to 10% of contract value and for performance can be up to 2.5% of turbine cost.

The company would like to work with a consortium which is led by a large Japanese company. The Japanese are known to the company but they have no previous experience of working together.

Risk Factor	Low	Moderate	High
Scores	0–3	4–6	7–10
Contract/commercial			
Conditions	Standard	Amended	Clients
Delay damages	Moderate	> 10%	> 25%
Performance damages	Moderate	> 10%	> 25%
Bonding	Default	Default	On demand
Fixed price	Variation	< 24 months	> 24 months
Ground risk	None	To assess	No assessment possible
Design responsibility	None	Shared	Shared
Defect liability period	< 12 months	12/24 months	24+ months
Exposure/Experience (weighted so that each score is multiplied by 5)			
Technology	Standard	Complicated	Difficult
Experience of partner	> 2 previous	2 previous	None
Geographic			
Access	Good transport	Poor transport	Difficult
Political stability	Stable	Disturbances	Corrupt
Language	English	EC	Non EC
Dispute recourse	English courts	EC	Non EC
Foreign exchange	None	> 25%	above 40%
Weather	Negligible	Moderate	Monsoon
Client			
Type of client	Experience	Known	Unknown
Project size (weighted so that each score is multiplied by 5)			
	< $15 m	$15 to 30 m	> $30 m
Scoring system	Low risk Up to 100	Medium risk 101–200	High risk 201–300

4 Key Factors in Operating and Sustaining a Business

The modes of operating and sustaining an international construction business are the responsibility of a range of people. In general, in a large organisation, the operations manager and the managing director will hold key responsibilities on many of the issues. However, it is a fluid situation which depends on personalities, the skills of individuals and their ability to work with each other. Before concentrating on the operating issues facing the team or individuals, it is worth placing the responsibilities within the context of the wider organisation. For the initial model there will be an emphasis on contracting-type organisations, purely because they allow study of a wider range of issues.

In general in a large organisation, for example, there are likely to be roles looking specifically at:

- Managing director – the overall view
- Operations – the practicalities of delivery
- Commercial – the finances of daily operation
- Marketing – the winning of future business
- Technical/estimating – the specialist expertise applied to tender or work
- Finance – the finances of the wider operation/long-term
- Administration/human resources – the critical people end of the business

In smaller organisations or in consultancies these roles are likely to be combined but the work still needs to be done. There are a variety of aspects to the operation of an international project or operation, which can be grouped into five key functions (although this is admittedly rather an arbitrary split): finance, on-site works, design, human resources and plant and supplies. We can look at each of these key functions to see who is responsible alongside the operations manager and look at examples of how the issues involved can be approached and the implications of this approach. The examples are anecdotal rather than steeped in theory and reference. As a result, contractor-based work is the major source of example, the result of my own experience. Remember, however, that neither responsibility nor approach is set in stone and other configurations may yield better results.

Finance involves both project specific finance and the wider operational requirements. It is obviously important to look at cash-flow and at the insurances and the costs of insurances. This is likely to be controlled in conjunction with the finance manager and the commercial manager, and it is a critical aspect of the role. This is studied in more depth in the second half of this chapter.

On-site works require basic control of the quality *versus* cost *versus* time. There are many texts on construction management, the core requirement of this function, and there is little point in revisiting this subject since many references cover the subject comprehensively. The only issue which can be lacking from these references is the sociocultural angle, which can affect how construction management is implemented on a particular site. This is studied in further depth throughout this chapter, although the case study below highlights the type of complex arrangement which can be involved.

Case Study 4.1: Hong Kong Experience

On a petrochemical installation project in Hong Kong, the owner was a Dutch-Anglo conglomerate, the main contractor was Japanese, the sub-contractors were generally local and sub-sub-contractors were often from the Chinese mainland. Cost and resource availability drove this complicated arrangement.

Comment

The cultural differences involved in this stratified approach to the project were numerous. The main contractor was initially involved in a constant cycle of dealing with time and cost at the expense of quality followed by periods of dealing with quality at the expense of time and cost. A number of staff were specifically employed to act as communicators across layers, an expensive addition to normal site operations.

Design management requires the control of cost, quality and payment and checking systems on the inputs that go to design specialists and the outputs that come back to a project. Again, this is the subject of many texts, many of them outside the bounds of construction and concentrating on procurement of services since design is classified as a service.

For a consultant with their own in-house design capability, the management of this aspect is a part of the core business, and consultants will be studied as case studies in Chapter 7. For a contractor involved in a design and build project the procurement of design services is likely to be done in conjunction with the technical manager and possibly the commercial manager. Most contractors will have procedures and processes to cover procurement and again there is little to be gained from re-inventing the wheel (Connaughton 1996).

One interesting angle to this is the isolation and management of substantial risk. Many contractors have design capability but there is sometimes a reluctance to bring this together with their contracting ability in major international projects. Anecdotal evidence suggests this is because there is a belief that splitting the risk is preferable to any efficiency gains from in-house activity. A problem with the work will leave the company with double the trouble if the work is kept in-house. In Japan, contractors have actually been prohibited from owning substantial design capability.

Human resources are key to the operation, with management falling between the operations manager and the administration/human resources. The management will be looking at the numbers and skills required for each project for both staff and labour, the skills required and the contracts and conditions. However, sociocultural factors are equally important, and these are studied later in the chapter. Getting the right people to the right place at the right time is an oft-quoted phrase in management texts, and again there are many good references for the theory. Implementation of human resource strategies for international construction projects often comes in two forms:

(1)   Where the company carefully guards control of the project, appointing locals or people of other nationalities at lower levels of management

(2)   Rapid localisation, where local staff are quickly recruited and promoted into positions of authority

The Japanese approach to management of these issues has been to stratify in an almost scientific manner (see Case Study 4.2). This is different from the Anglo-Saxon approach, where the argument is that in a global world maximum benefit is derived from mixing nationalities right through to board-room level. This second approach dictates the policy of UK and USA companies developing staff of all nationalities, a policy further explained in Case Study 7.2 in Chapter 7.

Case Study 4.2: Japanese site in Singapore

One site that I visited in Singapore had a Singaporean client, Japanese management, European and Australian engineering staff, Thai skilled trades such as steel-fixers, Philippine labour and a major item of plant imported from Russia. The belief in this extreme case was that time, cost and quality considerations were best fulfilled by splitting the work across nationalities. Management control is facilitated by having trades grouped by nationalities. The primary driver was, of course, costs since league tables of labour costs are regularly developed and showed at that time, for example, that Australian engineers were cheaper than Japanese and Singaporean or that Philippine labour was the cheapest, easily available in Asia.

Comment

This approach may have its critics among the politically correct, but it has support from a variety of sources. Anecdotal evidence would suggest that German and French companies follow similar approaches (ECI 1997).

Japanese research by, for example, Takaiwa (1985) has often meticulously translated selected field evidence into productivity calculations for international construction. More importantly, there are many studies which report on variations in productivity across nations, and these are often used in investment decisions in other sectors beyond construction, such that it is a widely used technique.

Plant and supplies also need management of cost, control, payment and input/output. A good example of this is the Hong Kong Airport which was outlined in Case Study 1.1 in Chapter One. The procurement of such a sizeable portion of the world's dredging capacity was the critical factor in the success of that project.

Case Study 4.3: Bakun Dam (Then 1996)

An interesting project of the extremes possible is the Bakun Dam, a major project in Malaysia which involved the clearing of forest and the building of a hydro-electric dam on one island from which electricity would be transferred by underwater cable to mainland Malaysia. By itself this appeared to be a massive but straightforward project which attracted the interest of a number of the global construction contractors, who hoped to lead a bid for the work. Closer inspection of the scope of the work, however, revealed that the underwater cabling was of a specification and length (665 km) that would take the combined production capacity of all the global cable manufacturers concentrating on large underwater cable, for a substantial period.

Comment

Thus, those who had been thought minor players, as suppliers of a specialist product, found themselves in a position to dictate terms to the main players. The project was eventually shelved after economic conditions deteriorated in Malaysia (Soong 2000). Despite the examples showing how every project is different, perhaps too different to standardise the response, there are common themes which good operations managers use to the full. In such a split of roles what is the operations manager likely to control or manage? One of the first questions is always likely to be 'Can we do the work alone?' In international work it is always likely to be an emphatic no.

The responsibilities outlined above are often very similar to the roles of an operations manager in domestic situations where there is often a tendency to look at a self-standing or closely-knit band of known players, since this greatly reduces risk. Even companies who would claim, for example, 70% of their work to be sub-contracted will be proud of their management of a tight list of trusted sub-contractors. International contractors (or consultants) would aim to reach a similar point where they are mimicking their domestic approach. However, in the early days of an international operation partners play a key role if the local presence is small. Most, if not all, of the responsibilities benefit from risk-sharing and this requires partners.

4.2 Partners

Why are partners necessary and how are they 'used'? There are, of course, a great number of reasons and chief among them on the international circuit is the desire to have local knowledge. However, they have a number of functions: eliminating competitors, mutual respect where they add genuine value to a team, for pre-qualification purposes, to suit local regulations, special technical requirements or costs.

Excellent references in the explanation of these various functions are Brandenburger and Nalebuff (1996) or Pietroforte (1996). There is thus little point in detailing these factors except to say that the general purpose of finding partners is to reduce or eliminate risk. Risk changes with time and the choice of partners can be equally time-dependent, i.e. there may be a need to have partners in different configurations at different stages of a project.

Prequalification, for example, is a critical part of the process despite the fact that it is rarely accorded serious resource in large organisations. Prequalification requirements, when they stray away from technical merits alone, can, for example, force the decision of which company becomes the lead partner. Political aspiration, such as the likes of Malaysia where quota systems have applied, can play a large part in the initial short-listing. A non-Malaysian organisation may thus need a Malaysian partner to lead prequalification for a major project (TradePort 2000) even if they are too small to lead the project itself and can then become redundant at the operations stage, or at worst become a passive, sleeping partner.

Case Study 4.4: Camisea project

One example of political aspiration driving the prequalification exercise was the ill-fated Camisea project in Peru, where the plan was to take oil from the Peruvian jungle via an Andean pipeline. The client, Shell, had a prequalification which demanded each contractor had a substantial local partner. The definition of substantial, however, in a continent where most players are small, left five local players for a short-list of five contractors and global giants were competing to 'win' these five as partners.

Comment

This is another example of small players becoming key to a major project. The project, however, is shelved at the time of writing since Shell have withdrawn (*Upstream* 1999).

Special strengths are the popular reason for seeking partners. The acquisition of a set of specific specialist skills or the scale to deal with massive projects often comes from partnering arrangements. Specialist strengths run in various forms, from simple language and customs knowledge through to specific engineering skills. A prime example of this is the French company Freyssinet, which specialises in pre-stressing operations. They have built a global reputation and, with few direct competitors, have a platform to command partnering or sub-contract arrangements across the world.

Cost is another important factor, although the maintenance of a lowest price globally in a competitive world such as construction is extremely difficult. Thus, the cheapest source of labour, supplies and staff varies with time, tracked as always by the quantity surveying industry. Quality is a related factor since cost is often traded against lower quality, although quality is much harder to measure.

An interesting example of known cost advantage has been Indian contractors. Many global players would like to take advantage of this, particularly across South East Asia, although the barriers are significant, primarily through immigration policy, since Indian cost-effectiveness depends on large numbers of manual labour.

What is the choice for a contractor, consultant or building material supplier in partnering roles internationally? This is not a decision to be taken lightly since partners can be a considerable source of problems and increased risk if the wrong choice is made. There are three types of partner available, as follows.

Partners as a term generally implies a risk-sharer, i.e. another player who will share risk and take an active role in the work. In any negotiation with a potential partner the share of risk is critical. Partnership arrangements are in two main modes: joint venture arrangement where each player has a specific % share of work, or a consortium agreement where each player takes a specific part of the work. The two can lead to an infinite variety, with any number of minor, major or equal partners. Most serious players, however, understand the benefit of simplicity and try to keep partnership to manageable, accountable proportions.

Advisers are often potentially non-risk-sharing, i.e. it would be difficult to pursue them for non-completion of a portion of the work. There are, however, an infinite variety and they have varying degrees of involvement ranging from absolutely no risk-sharing through to active participation and a risk-sharing partner in all but name. Again, advisers are employed in a variety of manners, from verbal agreement through letter of appointment to formal contract arrangement.

In the international construction market bogus advisers are a source of concern. Corruption can and does occur and corrupt officials often slip into the system under the guise of adviser. This is discussed in more detail in section 4.4.2 later in the chapter.

Suppliers and sub-contractors are a direct means of off-loading specific areas of risk and are seldom viewed as sharing risk with the main players. There is little difference between domestic and international projects in this sense. Critical to this is the correct ring-fencing of the supplier's role. If this is completed correctly then the main players can ignore that portion of work and concentrate on their own risks. However, there must be a balance between the cost *versus* quality *versus* monopoly. A supplier who dominates the main player through use of a monopoly can cause many problems and, in this instance, partnership would be a better option (see Chapter 3).

How are potential partners found? The PARTS framework (see

section 3.2.3) provides useful clues. In general, it is through talking to players in the field: through potential competitors, potential suppliers and buyers and related companies. Reading the relevant trade journals, reports from embassies, reports on the market or recommendations through third parties can all help. The actual method of approach can vary from discussion purely through third parties to the more direct approach of sitting down and negotiating. Culture, tradition and the level of previous contact all play their part in the choice.

4.3   Security and assurance measures

In the initial or pre-stages of a project the operations manager, together with the commercial manager, will spend much time on assessing the company's needs for assurance and security. The first stage, identifying the risks which may arise and the potential partners available, has already been explained in this and the previous chapter. The choice of construction arrangements can be the next stage, if these have not already been decided by the client. The projects themselves can run from traditional supply contracts for material producers, design for consultants or build for contractors through to packages of procure, design and build or even finance, design and build in the international market.

The choosing of suitable partners alone is not enough. The focus for much of this assessment will be the contracts which arise from the construction arrangements and the insurance measures available as a consequence, and the conditions they then impose on the company. There are two key elements to this: the financial status and the legal status of the construction arrangements and the obligations which arise, and attitudes to contracts and contractual arrangements vary enormously internationally.

Standard contractual arrangements range from simple memorandums of understanding through simple letters of specification through to standard contract forms (Freshfields 1996). Standard contracts for procurement offer a perceived advantage in terms of familiar responsibilities and risks. However, they are often amended and can suffer problems through translation into different national laws.

The key provisions in these contracts include the definition of scope of work, payment provision, design development, delay alleviation, dispute resolution and choice of law (Freshfields 1996).

All of these have important implications for an international project.

One of the greatest causes of concern often revolves around the dispute mechanism and the choice of location if a dispute should arise. The unofficial but golden rule for most companies appears to be to seek a neutral venue if it is possible.

What is needed from contracts and indeed other forms of assurance or security? Fair treatment is the major concern. The more standard the form, the more neutral the source of the contract form. The more detailed (but not prescriptive) the specification, the less chance of dispute. The more unconventional the contract and specification, the more risky it becomes.

Case Study 4.5: Experience in Taiwan

One example of this was a project, in Taipei, which involved the building of a tunnel in a design and build package. The proposed contract was a standard form for Taiwan and was compared to the usual conditions for standard international format. The appraisal revealed 75 changes to general international standard forms. Many of these were five-word changes which required detailed assessment of cost and risk implications.

Comment

Since this was a technically risky project before the addition of contractual risk any thoughts of a bid were dropped. The whole exercise was, however, useful since it led to a stronger understanding of the market.

Other forms of assurance or comfort come from setting up insurance arrangements or from knowing that the financing and payment provisions of potential projects are secure. The principal forms of insurance are through Export Credit arrangements or through privately obtained insurance to cover the usual risks of third party damage or injury, equipment damage or breakdown and contractors' all-risk or consultants' professional indemnity. Again there are good explanations of these provided by official government guides, the banks and the law firms.

Secure funding is a huge topic in itself. Often the simple question 'where does the finance come from?' quickly establishes how good a project is likely to be. Blue chip companies paying for their own development, state funded projects, development banks or bilateral aid packages are viewed as being fairly secure. Less secure are the arrangements where the aid or funding package only covers part of the cost leaving the company to find ways to fill the gap, or where the company itself has to find the finance initially to cover the project costs. This may entail the company needing to arrange a limited or non-recourse loan, where a lender will agree to lend the money and the terms of repayment are linked to the success of the project. Generally lenders will initially be uncomfortable with this level of risk and will seek more secure terms, although there may be grant packages from government to support the company itself rather than the project.

4.4 Social and cultural issues

This book so far has assumed that similarities in construction, as an industry across the globe, lead to generic approaches and standardised tools for interpreting the information. It is important, however, to look at the possible differences which occur across countries or projects. Much of this revolves around the human factor. Many of the differences which arise between domestic and international work come from social and cultural issues arising from the different traditions of peoples and nations or regions where the work is located.

Social and cultural issues manifest themselves in a variety of ways and it is important not to generalise in a manner that can lead to mistakes. At the same time it is important to recognise that people are individuals, and the difference between two individuals of the same nation can be bigger than the assumed difference between one national trait and another. The advice of many is to keep an open mind, focus on the common goals, develop a basic understanding of other cultures and not to assume that your domestic experience is the only right approach (McWilliam & O'Reilly 2000).

Case Study 4.6: Diplomacy in Japan (1)

An example based on my own experience occurred in Japan on a prestige visit with some junior politicians from the UK. The objective was to introduce four future leaders drawn from politics, journalism and business to the experience of Japan. Everybody except myself was new to the country and, as such, we were hosted by the British Ambassador on the first night. He explained the rights and wrongs of etiquette and concentrated for a while on the handling of namecards, with the usual official guidance not to write on namecards since it was taboo and showed extreme disrespect.

On the next day our first meeting was with some very senior Japanese civil servants. We exchanged namecards and somebody made comment that I had studied in Japan. Since this was unusual one of the Japanese wrote this on my namecard. The British delegation was horrified until I laughed (I had been there before and it is not that uncommon).

Comment

Guidance on the subject tends to be cautious for obvious reasons, and it is better to be careful than to cause offence and lose work as a result. The ambassador was correct in being cautious. At the same time other experiences in Japan revealed differences where guidance is seldom given. However, uncomfortable situations can arise.

Case Study 4.7: Experience in Japan (2)

When working in Japan I once made a mistake with a design project by badly underestimating the time that it would take to complete the project. There was a tight timetable, causing difficulties for the client on follow-on work.

The unhappy client summoned two bosses and myself for a meeting. Rather than threatening financial or legal retribution as one would expect in the West, the client called for an apology Japanese style. We therefore had to get down on our knees and apologise from floor level. The ritual humiliation was an episode that I will not forget, particularly the concept of my boss being humiliated at my expense.

Neither of the two above examples are major cases, and there is little point in trying to cover all the differences likely to be encountered across the globe since inexpensive sources of guidance are both numerous and extensive (my own experience has been through the British Embassies across the globe).

However, it is worth briefly reviewing the theory of socio-cultural differences since it can provide an insight into why the differences occur and this type of information is often not available through construction circles. Differences arise as a result of many obvious as well as hidden factors.

For example, companies have two main options for doing business across national boundaries: they can bring their own workers (where laws permit) or they can employ local staff. The human effects of this choice can be very different; for the first option there is a need to train the non-native in the accepted practices and allow attitudinal adjustments for the new country, the second requires careful selection, recruitment and training and the need to adapt company systems to suit. All of these require research and cross-cultural training, since people with different cultural backgrounds can misunderstand each other's behaviour, leading to interaction problems which frequently manifest themselves through a costly high turnover of staff (Berry & Houston 1993). Actual theory in this field is complicated and often derived from political considerations of human behaviour. It does appear, however, that the basic building blocks are agreed: that culture defines accepted ways of behaviour for members of particular groups and socialisation is the process by which individuals are then introduced into the culture of a society (Haralambos & Holborn 1991). Beyond this point, human behaviour is extremely complex and there are many ways to try to explain and generalise, and as such it can become a highly contentious issue. One of the simpler explanations (Hawkesworth 1994) suggests that the following are key dimensions in explaining the major causes of the differences.

4.4.1 Cultural factors

Power distance – Humans are very sensitive to their position within a hierarchy, although this varies enormously across different cultures. One indicator which is often quoted is the power distance from the person at the top of the company to the person at the bottom of the company, an indication of the social stratification, flexibility or even just shared values and attitude towards equality.

It is believed, for example, that power distance is very small in the USA whereas Japan is a highly hierarchical society. This generalisation, however, hides changing values with time and other indicators such as, for example, salaries which have a much flatter profile in Japan than the USA.

Uncertainty avoidance – The collective attitude to uncertainty or risk is another important consideration. The concept of 'face' in China and, to a lesser extent, in Japan is often quoted as affecting behaviour of many within these groups. It can become engrained in regulation with, for example, the entrenchment of life-time employment in countries such as India and Japan or restrictions on dismissal as in France.

Individualism versus *collectivism* – This is based around the belief that an individual's rights are more important than the collective's. The obvious extremes were the cold war attitudes prevalent in Western democracy versus the collective approach prevalent in the former USSR. However, even within Western Europe, there are the social democracy approaches of Scandinavia and the Anglo-Saxon individualistic approach of the UK.

Masculinity versus *femininity* – This is often mistakenly restricted to the attitudes on sexual equality. However, although equality is an issue the theory appears to take it deeper into contrasts between patriarchal societies such as Ireland and matriarchal societies often quoted to include Italy, and the ensuing differences which develop.

4.4.2 Socialisation factors

Regulation – The process of incorporating cultural values and norms in government supported legislation and regulation. This is the marking down of a value consensus and, by implication, agreement of what is regarded as deviation. This is a particularly political factor since there is a strong argument that a value consensus is rarely universal across a nation or culture, but is, in fact, the value of a small but strong leading group such as, for example, the Anglo-Saxon tradition in multicultural USA (Haralambos & Holborn 1991). The most obvious manifestation of this is through free-market versus socialist societies.

Tradition – An infinitely more easily understood phenomenon is tradition which becomes engrained in society to the point where it is accepted as marking one society as different from another. A good example of this is the Spanish siesta.

Religion – Again, a clearly identifiable factor in the social development of individuals and their attitudes. A useful example is in the contrasts between Christian and Islamic societies.

Having outlined the contentious basics it is obviously important to outline how these affect construction as an industry and as a place of work. However, since much of this will be anecdotal (good references are hard to find) two general case studies are outlined below, indicating how business cultures differ across Europe, before looking specifically at construction.

Case Study 4.8: Cross-European attitudes (Stanton 1996)

An interesting study of European attitudes to human resource management was conducted by Coopers and Lybrand across 14 countries and 21 industries. The aim of the study was to look at the convergence or divergence of thinking across countries, and whether the concept of a 'Euromanager' was a practical goal.

On questions of organisational hierarchy Swedish perceptions sit at one extreme while Italian perceptions sit at the other; most Swedes (and Americans) are happy to by-pass the organisation hierarchy if they believe it will be more efficient. Italians, however, saw by-passing their managers as unacceptable.

Anglo-Saxon cultures are keen to separate person and job, whereas Latin cultures see the two as inseparable since personal relationships are key to success.

Regulations which affect the value of pay comparisons, taxes, company cars, pensions and other benefits are significant differences across Europe. In this respect, the legislation covering industrial relations is a particular minefield of differences.

Finally, it was noted that leadership and matching strategy to corporate and national cultures (see Case Study 4.9 below) were key factors which presented special challenges across Europe.

Comment

The building of a single European market will be a major challenge, which goes way beyond changing rules and regulations. Converting national mind-sets, values and traditions will be difficult to achieve, if the desire is for a 'one size fits all'. One of the most contentious areas is attitudes to leadership.

Case Study 4.9: Leadership (Berry & Houston 1993)

Attitudes to leadership vary enormously. In some countries workers prefer participative management, where they share an element of decision-making with managers. In other nations there is a negative view of this and it is thought to be a sign of weak management, an inability to take responsibility. Studies in the 1960s showed Norwegians and Puerto Ricans to react negatively to the idea of participative management (although it is likely that Norwegians would show different results now). More recent studies show the Dutch to be advocates of the approach. Indian workers, by contrast, show no response to either participative or authoritarian approaches, and their attitude hinges more on 'caring'.

Comment

The obvious conclusion is that 'one size fits all' does not appear to be an option internationally.

How do socio-cultural factors show themselves in construction?

As noted above, the industry has not been flooded with socio-cultural studies of attitudes. There are undoubtedly some studies of socio-cultural aspects of construction (the author is aware of one on-going research study in the UK) although cross-cultural study is rare. As such the examples below give some flavour of differences, although they are neither rigorous nor well-founded in theory. Most importantly, there is very little option to comment on best and worst; each case needs to be managed to become an advantage rather than viewed as a positive or negative case.

Site practice

Site practice can vary enormously across the globe. Attitudes to health and safety (Case Study 8.4 in Chapter 8), productivity (Case Study 4.2 in this chapter) and the use of different materials such as bamboo *versus* steel scaffolding are frequently quoted differences, although they represent the tip of the iceberg. While some of the thinking behind these differences is undoubtedly economic or availability, there are areas which arise from socioeconomic conditioning rather than hard fact.

Advice would generally suggest the best option is to seek local advice and participation in planning in a specific region.

Management and respect

One generalisation which has reared its head so far in the case studies is the obvious differences between Anglo-Saxon and Asian approaches to management. These, however, go well beyond construction projects.

The roles of men and women

This is an interesting area for construction sites. In the Western world construction sites have long been the preserve of men, although slowly the barriers are breaking down and women are moving into mainly management roles on sites, as the concept of equal opportunity surfaces. However, in Asia where equal opportunity often remains a dream, women have long been seen on site, though mainly as unskilled labour.

Contract attitudes

This is perhaps best illustrated by a case study.

Case Study 4.10: Hong Kong experience

In 1998 a major investigation into irregularities and poor piling work on a building site revealed a complex web of contractual arrangements (Jones 1998). The client was Hong Kong's Building Department. The structural engineer was a UK firm. The main contractor was Japanese. The piling work was sub-contracted to a French company who in turn sub-contracted it to a German contractor. This was further sub-contracted to a local company who actually installed the piles.

Comment

Although typical for Hong Kong, there are few other countries where the concept of sub-sub-sub-contracts would be normal. By way of contrast Japan has had a very lax attitude to written contracts, often relying on verbal agreement and trust, frequently on a one-to-one basis, so that contacts are critical.

Attitudes to corruption and bribery

Corruption and bribery are very emotive words in the developed world markets. Much of the advice on overcoming cultural differences centres on keeping an open mind and not seeing different ways of doing things as being absolutely right or wrong, since cultures and norms vary enormously (McWilliam & O'Reilly 2000).

Thus, while corruption was not tolerated in the developed world (in theory at least) the attitude was different when working outside home markets. For many years the payment of bribes by the international arms of companies to corrupt local officials in pursuit of international projects was condoned by many governments, even though this was banned within their home countries. Viewed as a cultural difference and unavoidable in the worst locations it was a tax-deductible expense in several European countries including Germany (*Economist* 1997). Besides, developed countries, from where many of the international companies originated, while relatively free of straightforward bribery, were guilty of other grey or murky practices: domestic cartels in, for example, France, Holland and Japan (see references in Chapter 5, section 5.2) fixing prices and competition or the tying of aid packages to dubious conditions for procurement (*Economist* 2000). The hindering of competition through such practices was therefore a well established if frowned upon practice in the developed world.

Thus, there can be little surprise that increasingly construction is being targeted as the world's most corrupt business, a distinction it currently shares with the arms trade (*NCE* 2000). Development Banks, such as the World Bank, now have a clear anti-corruption policy which is applied to their projects although this is a relatively new phenomenon and probably still suffers from local handling of procurement in the worst locations. Corruption and bribery are more clearly viewed as a criminal activity and increasingly construction companies are being caught up in investigation.

Case Study 4.11: Lesotho Highlands hydro-scheme (Ungoed-Thomas 2000)

At the time of writing an interesting case was passing through the High Courts of Lesotho, a small country in Southern Africa. It centred around an $0.8 bn prestige hydro-electric scheme including a new dam and water systems to transfer water from Lesotho to South Africa, which has received much praise for its engineering content.

In the court case it is suggested that the dam's chief executive allegedly took over $1.5 m in bribes from an international consortium which included a major UK contractor and a major UK consultant. The money was apparently paid into a Swiss bank account in order to gain contracts. The court has still to pass judgement.

There have been many studies on the effects of corruption on a country and on the attitudes of businesses. Surprisingly, it is not the amount that is involved but the manner in which it is organised (*Economist* 1994) which causes most problems. In countries such as Japan or Malaysia where the problem is reputedly widespread but the system is efficient, the certainty involved in having almost set percentages (see Chapter 10, section 10.4) removed by one designated official from contracts appears to cause few problems, although others would argue that the money could be better spent in a more honest fashion. Greater damage occurs for countries and concern arises for business when there is uncertainty as to who or how much is involved.

Case Study 4.12: The Russian system

In articles on bribery or corruption Russia often features prominently. One article on the subject (*Economist* 1994) describes how Russia had moved from an organised system run by the communists to a disorganised one with no discipline. In the early 1990s opening up a business required bribing local and central government officials, fire and water authorities, police and the local mafia. All charged the highest rate they could get with no reference to the other groups. Thus, it was impossible to predict the 'going rate', frightening off many potential investors.

Transparency International is an organisation which tracks the extent and nature of corruption across the world. Its website is a useful source of information. It produces an annual corruption league table of the world's best and worst (Transparency International 1999), measured from surveys rather than in absolute or quantifiable terms. In 1997 the world's worst five countries were:

- Nigeria
- Bolivia
- Colombia
- Russia
- Pakistan

In fact, the most interesting thing of note from the above list is that the countries are so diverse in geographical terms, indicating that it is a global phenomenon and not restricted to one particular culture or location. It is often stated that work in one of these locations will invariably require a knowledge of how the 'system' works.

4.5 Estimating

Continuing along the theme of differences, the cost of work is often highlighted as the major difference between working in various locations. While it is true that costs vary across nations it is also true that they vary within nations. It is useful therefore to concentrate on the costs which may be directly attributable solely to international work. Climate, differing standards of quality, different standards arising from regulation and other quirks of man in combination with nature can all cloud the issue.

Although the management of costs is an on-going process throughout the life of the project or business this section will concentrate on the initial phase: the estimation of costs of work. Although not a direct responsibility of the operations manager the estimating function is vital to the success of the operation. In the UK where estimating was often separated, increasingly operations staff are playing a bigger role in this side of the business. In other parts of the globe such as Japan there has always been a belief that operations staff should be involved or should lead the tendering process.

Good estimation is vital to the operations end of the business since it is important to get the cost estimates as realistic as possible. Despite this, estimation can still be a fairly inaccurate business. It has been reported in the UK domestic market that estimating averages at best 24% accuracy (Waboso 1996). Given the additional complexities of international work it is vital that great care is taken in establishing a best estimate.

The best source of estimating information is, of course, local knowledge from knowledgeable sources. In international work this equates to information direct from experienced players in the country or region of the country concerned. This adds further support to the idea of working closely with local partners or establishing a strong local presence on the ground from the earliest possible moment. The typical project cycle of finding out about projects, prequalification and tendering lends itself to a steady build-up of relationships which can lead an international player to establish itself locally and benefit fully from local contacts and knowledge. However, there are many occasions when this is not possible.

Where the application of the best local knowledge is not possible, then there are other, riskier methods of developing a cost estimate. Time constraints are often one of the biggest problems for gathering all the best local information for a good estimate, especially since confidentiality can often impose further restrictions.

There are a number of specialist quantity surveying firms who provide a service monitoring international prices and their fluctuations across countries and with time. Their ability to supply knowledge and information can be extremely useful.

Market testing experiments provide another source. This applies as much to the construction market as to others (at the time of writing the fashion in the UK is to compare car and supermarket prices). For studies which are related to the construction market this can represent useful information for the budding international estimator, although it is, of course, an infinitely riskier prospect. One such study was conducted by Davis Langdon Consultancy on behalf of the DETR (Meikle *et al.* 1998). The work involved a comparative study across Holland, Ireland and the UK using firms of quantity surveyors in the three countries. A number of building types were analysed, among them a light industrial building. A building specification was drawn up assuming a multinational client wishing to develop a fairly standard type of building. The generic building type – a single storey steel portal frame with metal cladding, and an accompanying bill of quantities were common across all three countries. Site conditions were chosen as optimal green-field with all services available, i.e. the site and topography should not have a significant effect on cost across the three countries. The analysis was developed using information collated from recently completed work by the three companies, but using standard local practice and procedures.

The study revealed substantial variations in the costs of elements of the buildings. This was the result of a number of factors: the pad foundations frequently employed in the UK and Ireland are not normal practice in Holland where piled foundations are the norm, and roof cladding was particularly expensive in the UK while plastic guttering was very inexpensive in Holland. Taxes, professional fees, etc. also varied considerably with preliminaries representing 10% in the UK and Ireland and 21% in Holland. VAT varied from 0% to 17.5% while professional fees varied from 6% to 16%.

The net effect, however, was total prices which varied only by 3% (although construction costs varied by 30%) when all were converted into one currency at the current rates of the time. It is important not to read too much into this limited analysis. Some of the differences may reflect misjudgements or estimates from one of the companies rather than real differences.

However, one factor did stand out of this work as being critical to differences. The initial work revealed the UK to be the most expensive for the base line date (third quarter of 1997). But it was also shown that if the base line was three months earlier, Holland was more expensive. Six months earlier than this both Holland and Ireland were more expensive than the UK. Thus, currency movements appear to be a bigger risk than any structural cost differences. This should not be surprising since many other sectors view currency fluctuation as the single most critical factor beyond guarantee of payment (see section 3.6). It also explains how so many of these comparative studies find such differing results (another example is included in Table 10.2, Chapter 10, section 10.2).

Davis Langdon and Everest (1995) also produce international cost estimates, published by Spon. These guides can be used as the basis for developing estimates (see problem solving exercise 3 at the end of this chapter). There are many risks inherent in completing estimates at such arm's length from the market, and it has to be asked whether a company which needs to resort to this has not, in fact, overstretched itself and is quoting or bidding for work that it will have extreme difficulty in completing satisfactorily. Such risks include:

Standards	– materials	Unknown local costs
	– sizes	Currency fluctuations
Climate		Time – duration
Earthquakes and other natural phenomena		– approvals
Bloodymindedness		Lack of design approval
Need to be detailed and not rough when it counts		Acceptance at face value?

In conclusion, this chapter has considered diverse areas of work which represent important concerns in the operation of an international construction business. It has approached them in the same way that business runs, coming at all angles and heading off in different directions.

Problem solving exercises

(1) Review an article on an international construction project. Identify the sociocultural differences from your domestic market. What are the theories behind them and the reasons?

(2) Search the website of Transparency International (address in references) and check where the central European countries of Czech Republic, Poland and Hungary are in the tables given there. What are the implications?

(3) Using the information below, develop a budget estimate in Guilders (3.22G = £1) for a light industrial building to be built in Holland in 1998 to a similar specification to the building described. Provide as much detail as possible but, where this is not possible, provide an alternative method of costing and an explanation of the methods used. Provide a summary of the risks involved in presenting such an estimate to a potential client (see Appendix, Hints and Model Answers for Problem Solving Exercises).

The project

Cost breakdown for similar project close to major city outside London (adapted from Meikle *et al.* 1998, courtesy of Davis Langdon Consultancy from a project funded by the DETR). The multinational client wishes to build an electronics assembly plant. The site is a reasonably clean green-field sufficient in size for the plant with all necessary services to hand. The 2900 m^2 single storey unit has a clear floor area for factory space and an office space of 180 m^2.

Elemental cost (£)

Substructure   85 129
Reinforced concrete pad foundations to steelwork: reinforced concrete slab, power floating (15 kN/m^2 loading).

Frame   98 592
Steel portal at 6 m centres, 6 m internal height at eaves.

Roof   257 665
Contractor designed steel sheeting and insulation sandwich 37 mm thick (roll joints, colour-coated). Overhanging eaves with hidden steel gutter, flat panel fascia and highlighted flashing trim.

External walls   93 381
Horizontally profiled composite steel sheeting and insulation
sandwich panel 37 mm thick. Anodised aluminium curtain wall
double glazing with horizontal pivot opening sections.

Windows and doors   107 324
Anodised aluminium framed glazed doors. Electrically operated
doors to loading bay and timber fire escape doors, colour coated.
Double glazed roof light.

Internal partitions   10 751
Blockwork partitions, plaster finished and painted in offices,
ceramic tiles in toilets.

Upper floor of office   17 251
Precast concrete planks over office and toilets loading 2.5 kN/m^2

Floor finishes   5038
Carpet tiles to the offices, ceramic tiles to toilets, sheet vinyl to
entrance area.

Ceiling finishes   cost included in 'services' below
Suspended ceiling system with mineral fibre acoustic tiles to offices and toilets.

Services   286 944
PVC cables and conduits, lighting, power, ventilation, heating,
water supply as local practice.

External works   89 014
Concrete paved car park and footpaths, tarmac access road,
drainage, external lighting.

Construction inflation rates in Holland

1994	1995	1996	1997	1998
2.6%	1.7%	2.5%	3.0%	2.6%

Hint: unless you have a better source of information you will need
a copy of *European Construction Costs* (Davis Langdon & Everest,
1995 published by Spon).

For a final check remember:

Approximate estimating (1994)
Light industrial building for owner occupier   1360 G/m^2

5 The Global Market and Competitive Advantage

5.1 Splitting up the global market

The viability of any international construction business needs to be viewed in a number of contexts: the growth and stability of the company's domestic market (which often continues to be an important influence), the growth and stability of the regions or foreign markets in which the company operates and the global construction business and its sustainability. To study this we initially return to look at the world in total and the spread and size of the markets. The first part of this chapter looks at geographical and business patterns, and in the second part the possible effects of the patterns on business planning are studied.

The map in Fig. 5.1 is a copy of one developed by Roger Flanagan and his colleagues at Reading University (Flanagan 1998). It was developed as part of an Institution of Civil Engineers project called 'Exportism' which, as part of its brief, looked at the global trends within the civil engineering industry (Thorn *et al.* 1997). This is, admittedly, a rather narrow definition and very UK focused. However, it represents a useful starting point in looking at the world patterns and flows. The first point must be that the patterns shown are localised when viewed in a global context, and, more importantly, we must remember that most construction is conducted by local players for local clients. Truly global players or even just international players are, in percentage terms, a very small community. Thorn *et al.* (1997) quote a report which suggested that the top 225 international contractors won a combined total of $126 bn worth of work. With a rough estimated world market of $2970 bn (based on 10% of world GDP at that time (IMF 2000)), this represents only 4% of the market.

Traditionally international construction has implied companies from developed countries carrying out work in developing country locations. However, this is changing and the situation has become more fluid; Brazilian and Indian companies, for example, are gaining reputations across the globe. Much of this change is happening within regional blocks; a good example is Europe, as highlighted in a section 5.2.1, or South East Asia where the bigger

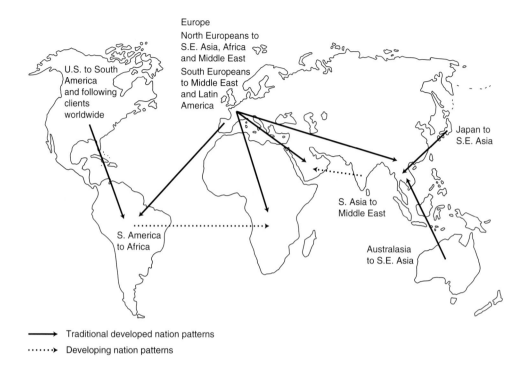

Fig. 5.1   World's construction trade flows (Adapted from Flanagan 1998).

national players from Malaysia, Singapore etc. are competing and investing in neighbouring countries (Bon & Crosthwaite 2000).

It was noted in Chapter 1 that the size of the world market is very difficult to measure and there is little agreement. For example, assuming construction represents 10% of global GDP the figure in Case Study 1.2 in Chapter 1 would produce $3200 bn. However, we have a construction derived figure also in Chapter 1 of $3600 bn for the global construction market. The difference, $400 bn, represents a value of work bigger than the combined construction markets of Europe's four big national markets. This illustrates the large degree of discrepancy available even within acceptable, knowledgeable sources. What is the value of attempting to develop a global estimate?

Chapter 3, section 3.2, has explained how, at the national market level, the measure of GDP is useful in assessing the size of the market, and changes in GDP are a measure of the growth and health of the particular market. These are important factors since a growing healthy market has greater potential than a declining, unstable market. Thus, attention tends to concentrate (together with competition) on the healthy, growing markets and the theory is that all players benefit from a stable growing global market. Tracking whether the global market is stable and growing therefore represents useful background work.

However, the growth and stability of the global construction market are fairly abstract since the market is the sum of many fragmented parts. It is also difficult to exploit, since the global product is extremely diverse, with service provision in some cases, manufactured product in others. The largest company in the world, Bechtel, had a reported international turnover of $10 bn in 1999, equivalent to 0.3% of the total and so small as to be insignificant in influence terms.

The reference to markets and market share arises from two interests:

(1) Defining an area of business where uniform rules, systems and behaviour are accepted, something that one company and its small group of senior managers can deal with and perhaps plan for through one business plan

(2) Assessing the ability of companies to influence behaviour and be proactive in the market rather than having no influence and, by extension, no ability to plan ahead.

Business management references often comment on the definition of a market and the difficulties associated with this. For example, does a company such as Mercedes form part of the general automobile market or are they a specialist in the luxury car market? The answer depends of course on what you want to do with the classification. Mercedes can be classified as being part of both and influenced by both.

Although the experience varies greatly across country and industrial sector, the general rule appears to be that a company's profitability can be affected by its market share. There is a very coarse rule that bigger market share can lead to bigger profit up to a ceiling of about 40% of market (Kotler 1997).

Where one company or small group of companies has a sizeable percentage of the total market it is widely believed that they can influence price changes, new product introductions, service provision and promotional intensity (Kotler 1997; *Economist* 1999), i.e. influence the general behaviour of the market. Porter's model (Chapter 3) would lead us in the same direction. The lesson therefore is that big is beautiful, and Mercedes is therefore quoted as being small in the general automobile market but clearly big enough in the smaller luxury market to make an impact.

At the other end of the scale the studies appear to indicate (although it is not explicit) that players with market shares of 10% or less will have little influence on the wider market. Thus, the international construction players with their combined 4% would struggle to drive the global construction market forward in a direction of their choosing. There may, however, come a time when the global giants reach a scale where they are truly global and start to influence the global market.

At national markets level, however, the market shares are closer to the size required to influence behaviour, and there are other factors in play (such as cartels, protectionism, etc.) which can give major construction players a considerable influence.

Globalisation has also created regional trading blocs encompassing more than one country, such as ASEAN, Mercosur, North American Free Trade Agreement (NAFTA) or the EU (IMF 2000). These are blocs where cross-country trading is an integral part of the business environment, supported in part by governmental agreements. The biggest three market blocs of the world, which show some form of coherence in terms of growth and stability, are probably the USA, EU and Japan, although two are clearly national markets.

These three markets share common internal features: they have common legal frameworks, common social frameworks and common economic values. In the case of the USA the legal framework is provided by federal and state legislation and institutions, in Japan it is national and prefectural and in Europe it is EU but still driven by national and local regions. There are, of course, many other local variations internal to all three markets through accepted standards, norms and traditions. Of the three, obviously the EU has the loosest common frameworks although clearly the trend within that market is for closer integration in all these features.

The size of the three big markets has been variously measured in 1994. Although these figures conflict with other references quoted (for reasons we have already explained in Chapter 1, section 2) they still serve as a useful first pointer. When compared to the global construction market estimates they show markets which potentially represent 26%, 24% and 18.5% of the total, very sizeable chunks of the global construction market.

With a combined total of over 65% of the world's construction market by value these three markets are likely to have a powerful influence on the overall market, although it is possibly in an indirect rather than direct manner. The market conditions of these developed markets will differ in many ways from other markets which are less well developed. However, there are trickle down effects through technology transfer, multinational client behaviour setting precedents, and evolving social and environmental pressures. The players who expand overseas will also add further linkages. We will look at the big three, EU, USA and Japan, in further detail later in the chapter (sections 5.2.1–5.2.3).

The biggest players in each of these markets (Table 5.1) are too small to dominate their respective home markets since their market shares, shown in the last column of Table 5.1, are insignificant. However, at the national market level in the EU bloc, the size of the leading players becomes significant with, for example, Holzmann having a market share of 7.9% of its home market in Germany.

Table 5.1   The big three markets in the world.

	(1)	(2)	% GDP	Biggest player
EU	$790 billion	$788 billion	10.80%	Holzmann 1.5% of market
USA	$710 billion	$570 billion	8.90%	Fluor Daniel 1.1%
Japan	$550 billion	$770 billion	17.80%	Shimizu 2.3%

Sources: (1) Flanagan 1998, (2) Davis Langdon and Everest 1995

If the market shares of the top five contractors are combined in each of the three global market blocs the Japanese players emerge as strongly competitive, as shown in Table 5.2.

Table 5.2   Market share of top five contractors in EU, Japan and USA.

	% of total market
EU	5.9
US	3.7
Japan	10.9

Can a bigger market than the traditional national market be defined for the construction market? Bon and Crosthwaite (2000) use blocs based on continents or parts of continents as regional blocs for some of their analysis of international construction. They suggest that there is sufficient linkage of countries within continental blocs to predict general patterns of future behaviour for these regional construction markets. On the basis of their analysis they conclude that Asia and, in particular, China will be the market for the future since the region's construction market is a bigger fraction of the world market than their national economies would imply.

Table 5.3 Contribution to GDP and global construction by continent.

	% of world GDP	% of world construction market
Asia	26	35
Africa	2	2
Europe	33	31
Latin America	7	8
North America	30	22
Middle East	2	2

Reprinted from Bon and Crosthwaite (2000) *The Future of International Construction* with the kind permission of Thomas Telford Ltd, London, 2000.

However, patterns of behaviour vary quite considerably at this level and it is therefore difficult to group countries in such big groupings. For example, Bon and Crosthwaite use other methods of classification to augment this initial grouping. Although defining market blocs is difficult, it is important to look at what we perceive to be the coherence which establishes our markets. For the purposes of this text we are trying to identify generic patterns upon which an understanding of the international market can be developed.

For example, as the case study in Chapter 1, section 1.4, showed, common legal and market frameworks are missing in South East Asia although there are other common trends such as an openness to foreign players and to projects developed from concept through to completion by one player. As Bon and Crosthwaite (2000) stated, South East Asia's countries share an exciting potential for the future, when measured in terms of fast construction spending growth and high construction volume, and many companies identify them as target markets.

5.2 Defining the important characteristics of national markets

National markets are clearly important, and the majority of information that provides the backbone of international construction is usually derived on a national basis. It is at this point that there is a strong temptation to describe a few key national markets as they exist today, particularly exciting markets of the future. Bon and Crosthwaite (2000), for example, suggest China, Mexico, Argentina and Egypt as prime potential markets. There is, however, a problem in that such explanations are likely to quickly fall out of date.

Case Study 5.1: The Asian crisis of 1997/98

In early 1997 South East Asia was still being hailed as the hottest economic spot on the globe, and had been described as the construction sector's most important international market for the last two decades (NCE 1999). Within a matter of months, an economic crisis led to a domino effect in falling confidence across the region, and stories of riots, melt-down and large numbers of players quitting the market. It was reported, for example, that over $30 bn of projects were postponed in Indonesia in one year (Bolton 1998). Throughout 1999 analysts spoke of SE Asia in terms of 'don't touch with a barge pole' warnings, and yet by the year 2000 (less than two years later) SE Asia was being hailed as the growth market for the future again (NCE 2000a).

Comment

The attitude of many commentators to the events which occurred within SE Asia was fickle as the response snowballed to embrace everything Asian, rather than specific problems. However, the event itself does underline the importance of the interconnections and confidence as part of the system.

In analysis we therefore need a broad generic framework rather than concentrating immediately on specific country issues, i.e. a look at common characteristics which have an important influence on business behaviour and then studying how construction companies deal with those characteristics.

The big three regional markets of the EU, Japan and the US are all commonly defined as 'developed' nations. By contrast, much of South East Asia, for example, is classified as developing or emerging nations. This three way categorisation of nations is central to the development of information for business use. It points to a set of risks which are different for each country, but which lead to a similar overall net effect. These net effects are, in a sense, then banded to allow us to look at risk and reward, as shown in Table 2.4 in Chapter 2.

To summarise, in theory at least, this leads the banding of developed (advanced industrialised) nations as low risk, low reward; developing (newly industrialised) countries as medium risk and reward; a new category of countries in transition as possibly slightly higher risk but medium reward; and emerging (less developed) nations as higher risk and reward.

Again it must be pointed out that this is a gross over-simplification, since different sectors of businesses have different risk profiles and rapid changes occur in risk and reward profile.

However, these simple classifications have become the basis for generic approaches which apply well beyond the bounds of the construction sector. The greatest advocates of this type of categorisation of nations are the supra-national bodies such as the World Bank or the OECD (Organisation for Economic Co-operation and Development), the nations themselves and securitisation industry (i.e. the insurers who have to pay when risk outweighs reward).

Official work in this field is conducted both by national governments and by bodies such as OECD (2000) or the World Bank. Official classification and the breakdown between developed, developing and emerging countries (among others) is conducted on a number of key indicators and different terms are used by many of the organisations. The OECD, for example, tends to concentrate on GDP and GNP per capita as the key indicators, the EU relies heavily on a mix of GDP and unemployment, and the UK government appears to set great store by disposable income. All of these bodies have detailed definitions of the differences between the three and they do not always match.

Case Study 5.2: The World Bank

A detailed analysis is conducted on each country where the World Bank operates, and much of this is made publicly available through publications and the website (World Bank 2000). The format of information on countries changes but a recent trawl through the site to look at information on Belarus revealed the following information available:

- A description of major projects currently planned or being implemented
- A brief description of the up-to-date state of the country and its politics
- A country profile highlighting all of the key social, economic and environmental indicators
- A 'country at a glance' document which provides details of the country's development needs
- A list of publications, recent press releases and contacts in the country

Comment

Apart from the obvious direct value of the information above, the categorisation of national markets provides important clues on many other aspects of the market.

The government statistics which accompany the work on categorisation are often viewed as 'hard' sources and are much liked by government and academics as being almost factual. Factual implies acceptability although it must be remembered that they are purely estimates since the nation's wealth is too complicated to measure with any accuracy. Japan's figures, for example, are notorious for requiring substantial revision over time.

The reasons for a government developing statistics are numerous; principally it is to enable them and others to assess their progress. However, there are two uses for commercial purposes; a rich or developed nation such as the UK bases much of its aid spending on the classification (an important part of international work for many players), and insurance premiums are often based around these classifications. Thus, similar information is used to classify aid and insurance packages. An example of this is shown here, a list from the UK of low income or emerging countries in 1993 (Morris *et al.* 1995):

China, Indonesia, India, Ghana, Lesotho, Pakistan, Philippines, Sri Lanka, Zimbabwe, Afghanistan, Angola, Bangladesh, Benin, Bhutan, Bolivia, Burkina Faso, Burundi, Central African Republic, Chad, Comoros, Djibouti, Egypt, Equatorial Guinea, Ethiopia, Gambia, Guinea, Guinea Bissou, Guyana, Haiti, Kampuchea, Kenya, Kiribati, Laos, Liberia, Madagascar, Malawi, Mali, Mauritania, Mozambique, Nepal, Nicaragua, Niger, Nigeria, Rwanda, Sao Tome, Senegal, Sierra Leone, Solomon Islands, Somalia, St Helena, Sudan, Tanzania, Togo, Tuvalu, Uganda, Vietnam, Yemen, Zaire and Zambia.

A similar list would be produced by, for example, equivalents in Germany, France or the USA. Clearly, however, classifications change with time. Bolivia in the above list, for example, had moved to middle income by 1995.

The contrast with private sector analysis lies in the type of information developed. In the private sector the concentration is on credit and risk assessment, which increasingly is requiring a well-rounded assessment of the national characteristics. Sources such as IMD, BERI or the *Economist* tend to combine the official hard statistics quoted above with softer information based on perception or confidence. These are often disliked by statisticians and academics as having little basis in fact. However, businessmen use them extensively for fairly simple reasons: the sources themselves are developing a reputation for being authoritative, the information is often simpler to understand, and perception or confidence play a large part in the business cycle of an economy (Santero & Westerlund 1976).

Case Study 5.3: IMD (1997)

The IMD yearbook is a very useful source of reference material. Each of the countries studied is ranked through analysis and survey of a number of factors:

Domestic economy   Internationalisation
Government   Finance

Infrastructure   Management
Science and technology   People

The 1996 yearbook ranks, for example, Russia at 46,45,46,46,46,46,31,41 in each of the factors out of 46, leaving the total as 46 out of 46.

Thus, there is a wealth of information available and it is important for a construction company to identify the most important pieces, review their possible implications and incorporate them in the business planning for the operation as a whole or for individual projects.

The pattern of developed, developing and emerging is most often translated directly to a risk–reward scenario as explained before. However, it is interesting to note the huge variation of interpretation that results. Table 5.4 shows a range of sources noting, for example, country risk, attractiveness, productivity or profit opportunity, but with little clear pattern.

To translate the generalised figures in Table 5.4 into construction terms further factors need to be considered. Bon and Crosthwaite (2000) have done an interesting analysis of global construction using the harder type of statistics as compiled by ENR and their own confidence/survey type of analysis. This is more rigorous than the explanation below which strives to explain the trends rather than provide academic correctness. Reference will be made to Bon and Crosthwaite's work, but it must be noted that the subject itself is complex.

Thus, in examining the information, it is important to allow for the following trends.

Aid *versus* public *versus* private funding

As nations evolve from emerging through developing and into developed, an interesting factor is the change in funding source for infrastructure: from a combination of substantial aid combined with private and public sector funding through to a cut-off of aid as private sector and public sector proportions of funding increase. The effect of aid should not be overestimated; currently, for example, it would represent 0.3% of the total funding source for Russia (OECD 2000). Anecdotally, it has been suggested that Ireland, another example, had 10% of its infrastructure paid for

Table 5.4 Measures of competitiveness.

Source	USA	Japan	UK	Germany	France	Italy	Russia	Brazil	Singapore	Measured quantity/quality
Economist 1997	8.1	6.9	8.3	7.9	7.8	6.3		6.5	8.2	business environment (out of 10)
Economist 1996							90	60	5	country risk (out of 100)
IMD 1997	58,028	79,151	43,251	67,419	70,629	55,408	5,437	7,338	41,816	productivity (GDP/person $)
BERI 1996	68	77	61	71	63	49	35	40	68	profit opportunity (out of 100)
Schwab et al., 1997	17,329	13,500	4,902	8,659	4,254	2,076	351	1,094	9,396	competitiveness (subjective ranking)
DoE 1996	95	137	118	107	122	117				construction productivity (GDP/person employed)
Davis Langdon and Everest 1995	5.3	1.9	1.1	2.4	1.7	1.5	0.3			construction output/capita (ECUs)

through EU aid packages. However, the presence of aid is of interest to many construction companies since it represents a stable money source, although it also has its disadvantages (see Chapter 9, section 9.3).

The graph in Fig. 5.2 shows the ideal, assuming little or no external effects. The real world adds the effects of business cycles, biases in government intervention policy and the vagaries of supply and demand, which are explained in the following sections.

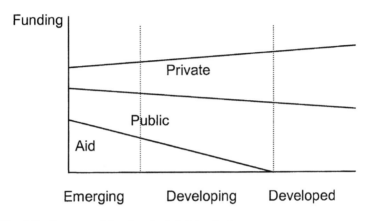

Fig. 5.2   Sources of funding for infrastructure.

The effect of business cycles

There are two business patterns which can affect the volume and type of work available, as illustrated in Fig. 5.3. The general business cycle of boom and bust is a well known short-term phenomenon which affects the whole economy of a nation. It is often believed that the construction equivalent is a slightly different shape to the general one, with construction entering recession earlier and leaving later than other industries. It is basically caused by a mismatch of supply and demand and therefore has a direct effect on the volume of work available, and an indirect effect on the profit or reward available.

Long-term a nation provides itself with infrastructure which should last for at least its design life (typically 30 to 100 years). As the assets grow the need to add should in theory decrease reducing the percentage of GDP allocated to construction. There are many caveats to this since assets need replacement, fashions change and technology moves on, all contributing to irregular patterns which make it difficult to detect the underlying pattern.

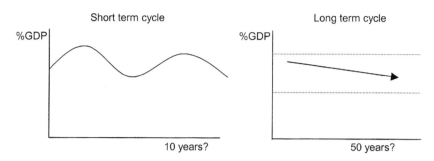

Fig. 5.3   Construction output trends.

Bon and Crosthwaite (2000) approach the above subjects from a different angle. Their belief is that, as nations evolve into developed status, the volume and type of work change. In the emerging stages construction spend as a percentage of the nation's GDP increases; in the developing stages it peaks before gradually declining in both volume and value as the developed stage is reached. In the earlier stages there is an emphasis on new infrastructure but this switches to repair and maintenance at the developed nation stage.

Bon and Crosthwaite believe that the short-term business cycle typically follows closely the more general national economic cycle, with references suggesting an average 16 year cycle. Although agreeing with most of this analysis, my own admittedly anecdotal evidence would suggest that rather than a continual decline, the percentage of GDP may settle at a lower rate (from possibly 10% to 11% towards 8%). Some of this may be classified outside typical construction classification. It is well-hidden, however, behind all the other complex cycles involved.

Supply *versus* demand

The management of supply and demand is an age-old problem. Many governments believe that it is their duty to try to flatten out the short-term business cycle (see Fig. 5.4) although a perfect match can seldom be made. Governments attempt to manage supply in order to maintain low unemployment, a stable market or the upkeep of prestige infrastructure but are frustrated by their inability to control or record all activity within a sector. Although they can spectacularly fail, the main aim is to flatten out the business cycle and keep it positive if possible. A prime example of

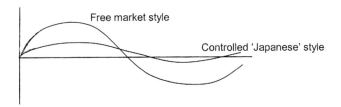

Fig. 5.4   Contrast of styles.

this has been Japanese government policy as illustrated later in section 5.2.2.

The net result of these various 'patterns' is, of course, an irregular summation which causes huge problems in interpretation. It needs to be added to the geographical patterns that were illustrated earlier, and makes the theory of companies using this to their advantage a complex one. However, companies can and do use this information to their advantage, often through choosing markets which are complementary and diverse in these patterns.

Fig. 5.5   What happens in practice – business cycles combined.

The random, immeasurable nature of this theory makes it difficult to see how a construction company can use this to plan ahead and become an international player. Sections 5.2.1 to 5.2.3, studies of the big three market blocs, provide some explanation of how the players and market characteristics interact.

5.2.1   The EU – a tension between national and transnational markets?

In the early 1990s Europe's pursuit of a single market became a centre of global attention. Players from the other two big markets,

Japan and the USA, established footholds in preparation for the event. European players established cross-border operations. However, all players experienced numerous difficulties exploiting the principle as national markets continued to show diversity.

Case Study 5.4: The Japanese view (RICE 1994)

The 1992 implementation of the Maastricht Agreement to promote free exchange of goods and services brought high expectations both within the European Union and to the many foreign players with a presence in the market. The Japanese, in particular, welcomed the liberalisation of procurement systems and carefully researched the market. They noted, for example, that 3500 construction projects were subject to international bidding under the new system. However, the reality of an open system was very different to the system on paper. They were, for example, unable to find out how many of the 3500 projects were contracted to foreign firms. Anecdotal evidence (Nakayama 1995) suggested that the informal networks necessary to build up a credible bid were closed or difficult to penetrate.

Comment

The evidence of difficulties did not stop big Japanese players from being optimistic about the changes, although they were aware of variation across the EU. The next case study illustrates changes which justify their optimism.

With a combined construction output of 860 bn ECU (1 ECU = $1.16) in 1994 the European market potentially represents the largest single construction market in the world. It is, however, a very diverse geographical area, made up of many major national markets. The European Union contains 15 diverse countries and possibly 75% of the total volume of European construction. Western Europe outside the European Union represents another 9% and Eastern and Central Europe provides about 16% of the total (Davis Langdon Consultancy 1996). These three geographical zones represent very different markets, with much variety internally. The markets of Eastern Europe, for example, will continue to draw interest for many external players since the area is underdeveloped and represents a large potential market sitting on the edge of a fairly sophisticated market (see Case Study 1.8 in Chapter 1). There are many areas of extreme poverty, however, and development may be a slow process. The building material producers and skilled and unskilled labour of the area represent a source of cheap supply for the more expensive markets of the EU.

The European Union is still in its infancy, although it is gradually assuming greater powers and responsibilities in the setting of legal frameworks and establishing common social and environmental standards. The traditions and rules of the national markets continue to dominate the daily existence of construction firms. Attitudes to technology, customers and even social and environmental pressures have varied enormously across the individual nations. Although the stated intention within the EU has been to create a single unified market the area still contains markets as diverse as Portugal, with a GDP/capita of 6500 ECU and a total construction market of 9 bn ECU, and Germany, with a GDP/capita of 17 800 ECU and a total construction market of 190 bn ECU (Davis, Langdon and Everest 1995).

The single market

As a result of the Maastricht Treaty and the many other EC directives incorporated into the domestic laws of each member country, the countries within the European Union have moved rapidly towards a single market. Many sectors of business quickly viewed Europe as a single market, with many of these benefiting from cross-national political goodwill or the huge economies of scale achievable in servicing a single market.

However, as the earlier Case Study 5.4 indicates, the liberalisation of the construction market has moved at a slower pace, as protectionism, tradition and the benefits of local knowledge created barriers to greater openness (RICE 1994). Construction continues to remain the sum of fragmented parts rather than a single, continuous market running to one economic or business cycle.

It is clear, however, that liberalisation has occurred (see Case Study 5.5 below). The EU now has a procurement policy which requires all public procurement projects of a certain size to be advertised across the 15 countries of the European Union. While this applies to all types of procurement, construction by its nature and size is heavily affected by the policy and most major construction players ensure that their sales teams scrutinise the official EU procurement journals on a weekly basis.

Case Study 5.5: UK example on CTRL contracts

Already the more open markets such as the UK are experiencing big changes. Section two of the Channel Tunnel Rail Link, which will extend the existing Channel Tunnel Rail Link from North Kent under the River Thames and into St Pancras Station in central London was put out to tender in 2000 (NCE 2000b). Some of the work, worth £600 m, was split into five major projects and short lists of three contractors invited for each project. The size and scope of work have created a desire for many players to seek the work in joint venture. Across the five contracts there are 21 companies recognised as UK based and a further 29 which would be recognised as European based.

Comment

For the Channel Tunnel itself there was a closed UK-French consortium bid. Despite the many obstacles to a single European market in the construction sector there is an increasing belief that the bigger players must be European in scale and spread. Large, transnational projects similar in size and scale to the Channel Tunnel have been developed by the EU to improve co-operation and infrastructure across Europe. The Trans-European Network Scheme (TENS) were 12 major road and rail transport projects identified by the European Commission as vital to Europe's cross-national infrastructure. It was envisaged that funding for the projects would be a combination of European Investment Bank, European Commission and private sector sources (King 1998). The case study below highlights a number of these projects on an illustrative journey across new Europe.

Case Study 5.6: New journeys across Europe by car (NCE 2000c)

It is now possible to travel in a car from Britain to Sweden. The last link in the network was the 27 km Oresund crossing. The journey starts in London streets and takes a motorway to the Channel Tunnel. A shuttle train will take car and driver through the tunnel to Calais without a passport check. New highways in France then take the driver through to Belgium's older motorway system. Holland, Germany and Denmark then follow, before reaching the Storebaelt crossing from mainland Denmark to Copenhagen. Oresund is a \$3.6 bn crossing with a cable stayed bridge built by a European consortium led by Swedish, French and UK contractors. The Swedish end boasts a new road system to take the driver into Malmo.

A driver can average over 100 km/hour for most of the motorway journey, much of it on dual or three-lane roads. Contrasting approaches to construction can be seen along the way: road quality varies significantly, the roadworks best in the UK, tolls at the Channel Tunnel, on France's motorways and at the Storebaelt and Oresund Crossings, good continental motorway services and parking arrangements vastly superior to the UK.

This 'Europeanisation' of projects, partnerships and funding is reflected in the increasing reporting of European league tables of the major building material producers, consultants and contractors. These often replace rather than complement the previous devotion to national lists. There are a number of issues raised by the last two columns of Table 5.5, which shows Europe's top ten building material producers. The period covered is clearly one of restructuring for the players as many of the names have changed, indicating a volatile, competitive market. The companies which appear to be most consistent are the cement manufacturers, who are described in more detail in Chapter 6.

The products represented in the Table are varied: brick, cement, glass, coatings, electrical products and cable specialists are all represented. It is noticeable, however, that large-scale construc-

Table 5.5   Top ten European Building material producers by international turnover.

Company	Country	Overseas turnover 1998 ($bn)	Overall position in Europe 1996	Overall position in Europe 1998
Saint Gobain	France	10.25	4	1
Holderbank	Switzerland	6.30	7	3
Schneider	France	5.77	5	2
Wolseley	UK	5.18	10	4
Lafarge	France	4.82	8	5
RMC Group	UK	4.20	9	6
CRH Group	Ireland	4.00	—	8
Hanson	UK	3.64	2	7
AKZO	Holland	3.47	—	10
Pilkington	UK	2.00	—	9

Adapted from *Building* 1996, 1998

tional steel is not included in these European lists (by contrast it features heavily in US lists).

Economies of scale and specialisation contribute to the individual companies being able to grow, an indication that our earlier note on market leadership in specialist markets (in section 5.1) is one particular strategy of note. Turnover outside of their domestic market is a major factor for all of these companies.

France and the UK are particularly well represented in the top 10 lists and the wider top 200 lists, although the wider lists show a good range of product specialists from across Europe. Lists of European contractors show similar retention of the same companies in the same period, although most companies have suffered. Scandinavian players, which have grown through acquisition, are well represented as are Germany, France and the UK (Table. 5.6).

A striking difference from the building materials lists is the much lower percentages for work abroad, with the bigger exporters including companies from small domestic markets (one of these is the subject of Case Study 8.2 in Chapter 8).

A European table for consultants is a more difficult concept, since the question of language causes difficulties. In the global lists it is noticeable that English-based countries feature highly, as English is the global lingua franca. However, Table 5.7 shows the top ten from the global lists. This produces a strong showing from Dutch companies, famous for their English language ability, as well as the expected UK presence. It is difficult, however, to judge how much of this is cross-national within Europe.

The European Union and other European bodies, many of them

Table 5.6   Top European contractors by international turnover.

Company	Country	Overseas turnover 1998 ($bn)	Overall position in Europe 1996	1998
Bouyges	France	5.04	1	1
Kvaerner	Norway	4.23	8	2
HBG	Holland	4.15	—	10
Skanska	Sweden	3.70	9	7
Hochtief	Germany	3.31	5	6
GTM	France	2.94	4	5
SGE	France	2.86	3	3
Holtzmann	Germany	2.85	2	4
Bilfinger+Berger	Germany	2.65	7	9
NCC	Sweden	1.95	—	14
Colas	France	1.26	15	11
AMEC	UK	1.18	14	12
Tarmac	UK	0.97	11	13
Balfour Beatty	UK	0.83	—	15
Eiffage	France	0.71	6	8

Adapted from *Building* 1996, 1998

collections of the major players indicated above, are engaged in other activity assisting the development of the single market such as the setting of Europe-wide standards for building, health and safety and workers' rights. This will undoubtedly help many of these big players and assist a level playing field. A tension remains, however, between the desires of Europe and its nation states.

Diversity across national markets

Beyond the centrally controlled areas of EU policy and European standard-setting, the domestic concerns of the national markets dominate the direction and setting of policy on construction. National policy and standards therefore remain a critical component of the European markets.

The economic cycles of the various nations vary quite considerably. A forecast in 1996 (*Building* 1996), for example, predicted recession in Germany and France, slight growth in Italy, stronger growth in the UK and a bumper year in Portugal with 9% growth of its civil engineering sector. Thus, consistency across the EU is not yet on the agenda.

Within the EU, the major markets are Germany, France, UK and Italy, although the analysis in Table 5.8 is restricted to the first

Table 5.7 The Top 10 European International Design Consultants 1994 and 1999

			1994 International turnover (US$000)	% of total				1999 International turnover (US$000)	% of total
1	Trafalgar House	UK	1,156	87	1	AMEC plc	UK	1,032	86
2	Nethconsult	Holland	471	83	5	Nethconsult	Holland	659	100
5	Fugro NV	Holland	328	86	8	Kvaerner plc	UK	589	53
7	NEDECO	Holland	255	100	9	Fugro NV	Holland	487	89
8	Jaakro Poyry	Finland	250	77	11	Arcadis	Holland	391	64
9	Heidemij	Holland	246	67	13	Jaakro Poyry	Finland	351	83
13	Bouyges	France	182	85	17	Ove Arup Partnership	UK	282	58
14	Mott MacDonald	UK	176	57	22	NEDECO	Holland	232	100
16	Phillip Holzmann	Germany	161	87	23	Maunsell	UK	222	87
17	Ove Arup	UK	155	55	26	Mott MacDonald	UK	175	52

No others in global top 50. Source *ENR* 24 July 1995

Source *ENR* 17 July 2000

Table 5.8   Germany, France and UK GDP statistics.

1 ECU = US$1.26	Germany	France	UK
GDP (bn ECU)	1401–1437	1069–1068	690–823
Construction Market (bn ECU)	149–190	118–100	60–61
% GDP	10.6–13.2	11.0–9.3	8.6–7.4
Private–public split	69:31	74:26	82:18

Source: in each row, the first number is from RICE 1994, the second number from Davis Langdon and Everest 1995

three. Each of these is characterised by having a largely established infrastructure but where repair, rebuild and maintenance are becoming increasingly important aspects of the market.

It is reported that national procurement systems are similar across the big four markets of the EU, with similar bidding arrangements, pre-qualification and tendering arrangements (RICE 1994). These, of course, will converge further as EU influence increases. Davis Langdon and Everest (1995) provides a useful overview of size, the make-up of work and general procurement arrangements for all the national markets across Europe.

While the official arrangements are important the unofficial traditions and networks are equally important. Although much of the EU is reported as being relatively corruption free (see the Transparency International league table at the end of section 4.4.2 in Chapter 4) there have been numerous reported scandals involving construction-wide cartels and monopolies across, for example, France, Holland (Grant 1992) and Italy. In the UK, specialist sectors have been suspected of running cartels.

Anecdotal evidence suggests that the example earlier (Case Study 5.5) of CTRL and its trans-European bidding list would be less possible in some of the national markets although clearly they are all opening up in ever greater degrees. Thus, construction appears to be less transparent than possibly other sectors across Europe and there would appear to be barriers to foreign competition beyond the simple 'local knowledge is best' type of difficulty. It is clearly not as deep or widespread or as visible as the Japanese system of *dango* (see towards the end of section 5.2.2).

Case Study 5.7: French road projects

In 1998 Lyon's urban council was ordered to close the $560 m Northern Ring-Road, a major toll-road around Lyon. The concession had been awarded to a French-led consortium in 1991 and had not been put out to competitive tender despite being covered by a European directive issued in 1989 (NCE 1998). In the same year, a French-led concession agreement for the $1.6 bn A86 autoroute west of Paris was also cancelled because it had not been awarded under open competition, as European law required.

Comment

Reports on French construction have frequently referred to issues such as this which involve French contractors from the largest such as Bouyges down to the smaller in a system which is clearly different from, for example, UK procurement systems.

In 1992 it was reported that Germany had 11 major contractors, France had 14, the UK had 10 and Italy had 2 (Davis Langdon and Everest 1995). Although scale varies considerably there is nevertheless an interesting story to these figures. By 1998 Germany had 9 major contractors, France had 11, the UK had 8 and Italy had 1 (*Building* 1998) as consolidation and competition reduced the number. Much of this was, however, driven by domestic consideration even though it often resulted in cross-border tie-ups (Morby 1996).

International or glorified domestic market?

The transnational market exists and is growing for many of the big players. It sits, however, alongside very local markets for many of the other players. Resistance and barriers still remain in the form of official law and unofficial practice.

In overseas markets the leading European contractors win substantial amounts of work with the combined total of the top four, Germany, France, UK and Holland, easily surpassing both the Americans and the Japanese. They were winning over $40 bn annually by 1998 (EIC 1999), much of this in the markets of South

East Asia. Again, however, close partnerships and relationships between the players supported by their governments have marked their progress. This sizeable export achievement represents over 10% of total turnover of their domestic markets, and follows a long tradition of targeting international markets.

5.2.2   Japan as a tightly managed national market?

In the early 1990s Japan was touted as the perfect construction market. Observers from around the world (for example, Sidwell *et al.* 1988; Lam 1991; Nahapiet 1995) came to study the well-managed construction market, the co-operation among competing players and clients and the heavy emphasis on innovation.

Note from Table 5.9 the differences in figures from earlier quoted estimates in section 5.1. Nature has left the Japanese with major earthquakes, a country where 70% of its land is classified as mountainous and without the resource of a strong raw material base. Historically, therefore, there has always been the need for Japan to manage these problems through selective use of materials and location and by choosing temporary, flexible structures and therefore cheaper structures over more permanent forms (which may not survive an earthquake in any case). Thus, the Japanese have seen themselves as being a fairly unusual island requiring different building solutions and standards.

Table 5.9   Japan *versus* EU *versus* USA market statistics.

Year	Japan	EU	USA
GDP ($bn)	4242	6645	6378
Construction (£bn)	770	748	471
% of GDP	18.1	11.3	7.4

Source RICE (1994)

World War II brought devastation, and Japan was left with millions of people homeless and a destroyed industry base. It fell under the influence of the USA, as the Americans set up and oversaw its new political system. The combined effect of an underdeveloped infrastructure and an opening up to new external ideas brought revolutionary change, which ultimately resulted in the country becoming the world's richest nation (Nahapiet 1995).

The provision of homes, the rebuilding of industrial structures, the replacement of temporary with an acceptance of more permanent forms of earthquake-proof structure all contributed to a long-term construction boom. By 1993/94 it was estimated that the market was worth $825 bn (Nahapiet 1995). Thus, a population of 120 million was generating the largest national construction market in the world (by comparison the UK market had a population of approaching 56 million (slightly less than half of the Japanese) but its market was estimated at $82 bn (roughly a tenth of the Japanese).

The might of the Japanese market was beginning to show globally, as the largest Japanese contractors became the top five global players by turnover (ENR 1995b), despite minimal overseas operations. The split of domestic work in Table 5.10 shows a fairly strong influence from the public sector, stronger than both the US and EU averages (27.5% and 39.8% respectively) at the time. This is a situation which allows government to take a fairly active role in management of the market as the huge amount of research, planning and analysis completed by the Japanese on themselves often reveals (RICE 1991; Ministry of Construction 1992).

Table 5.10 Domestic construction investment in Japan (RICE 1994).

Public (44.2% of total)		Private (55.8%)	
Non-residential	6.9	Residential	28.5
Civil engineering	35.4	Non-residential	17.8
		Civil engineering	9.5

The well-managed construction market

Wolferen (1989) provides a useful overview of the Japanese construction market and the interaction between the main government ministries, the contractors and the politicians. Close relationships remain the norm in Japan, with preparation and the development of a consensus approach to major projects a critical part of the process. Thus, group decision-making plays an important part in the process, a phenomenon which has been well documented beyond the narrow confines of construction.

Long-standing relationships have been the norm, with repeat business an essential part of the system, and cross-shareholding happening between clients and contractors and sub-contractors. The close bonds developed into families of companies known as *keiretsu*, which obviously changed patterns of power and control from those expected in other societies. The Ministry of Construction is reported to have had the largest of all ministerial budgets (Wolferen 1989). By 1986 the ministry was directly or indirectly managing over half a million public works projects through a network of 880 offices nationwide. Wolferen also reports that the movement of retiring senior civil servants to contractors was an important part of the system, extending relationships, stability for the post-holder and passing knowledge through the system. All of this added further to the links between leading players in the sector and the organised, controlled nature of the market.

Much of the work, as elsewhere, is provided through large-scale super projects, but ones which are specifically designed to be leading edge (OCAJI 1991). Recent examples include the Tokyo Trans-Bay Expressway and the massive development of Tokyo-Yokohama waterfronts, Kansai Airport on a man-made island and the completion of fixed tunnels or bridges linking all of Japan's four main islands. Much of the drive for the recent construction boom has been to provide infrastructure for the country's import/export of goods and the increased travel of individuals.

Case Study 5.8: Kansai Airport (Dilley & Taga 1995)

Kansai International Airport was opened in 1994 as the world's biggest airport terminal. It took six years to build, cost $15 bn and had 8000 workers on site at the peak period of work. The impressive terminal has an equally impressive foundation: a 511 hectare, 180 million m^3 artificial island constructed 5 km offshore in an average water depth of 18.5 m. The design has had to allow for earthquake and for sand and clay seabed settlements of up to 1.5 m during construction and a further 2 m over the following 50 years. The airport is accessed by a 4 km artificial causeway.

Although the reclamation project was a largely Japanese affair the terminal contracts were opened up to foreign competition, a relatively new concept in Japan at that time. It has been reported (*Asahi Evening News* 1989a) that this resulted in higher prices in order to progress the opening up of the market.

An international architectural design competition for the terminal building was won by an Italian architect, with much of the detailed design by a group of French, UK and Japanese con-

Fig. 5.6 Kansai Airport, the subject of Case Studies 5.8 and 7.3, was constructed in Osaka Bay, Japan. The massive artificial island was created using dredged sand before the airport and buildings could be developed. (Photos courtesy of Penta Ocean Construction Ltd, Tokyo, Japan)

sultants. Again, the critical design loads are earthquake design. Construction of the terminal was a three year project with many of the major building elements procured from overseas.

Comment

The cost and scale of work incurred from locating an airport 5 km offshore would be prohibitive in many countries, and many would question the need for the location. Kansai was one of seven major projects (later increased to 17) across Japan which were designated as special projects open to foreign competition. In keeping with the Japanese approach to most issues, the opening up of the market (see Case Study 5.9) was a meticulously managed process with higher prices a consequence that all accepted.

It is reported that in times of recession, as private sector investment dips, public sector investment rises to maintain a stable market (Nahapiet 1995, RICE 1994), i.e. the volume of publicly funded works was managed to suit the highs and lows of private sector investment. Although many countries make some attempt to do this there are very few who are so active and meticulous in pursuing such a policy.

This policy has continued throughout the later part of the 1990s as Japan went through a continued recession, and money was pumped into the construction market to keep it afloat (NCE 2000d).

The methods of procurement, working practices on site and employment systems are reported as being very similar to those elsewhere (GCAO 1993). Surveys report contractors' concerns about unrealistic deadlines, skills shortages, productivity and aging in the work-force, problems which occur across the developed world. These generalised issues, however, hide a few clear differences, caused chiefly by the very close relationship between all of the players and the apparently tightly controlled nature of the market. It has often been suggested that while the USA and Europe rely on contracts, prices, cost and competition, the Japanese business relationship is set on continuing trust and co-operation (RICE 1991).

Case Study 5.9: Attitudes of the Americans

The USA has played a very special role in the opening up of the Japanese construction market. In the 1980s tension grew between the USA and Japan as the Americans complained about the US–Japan trade imbalance and lack of openness of the Japanese economy. Initially USA action concentrated on high-tech products (*Asahi Evening News* 1989b) before spilling over into a wider field, to encompass construction.

When the two sides got together, an agreement – the US–Japan Major Project Arrangement – was signed in 1988 opening up 17 major projects to international competition and promising better access for foreign companies to the Japanese market. This entailed helping foreign companies overcome the many barriers to receiving construction licences, and by April 1989 an initial batch of eight

American companies, six Korean and one French had been granted licences (*Asahi Evening News* 1989a).

In 1989 a further problem arose when 70 Japanese firms were caught rigging bids for the contracts at the USA Navy's Yokosuka base. Of these 14 were then banned from future contracts with the US government. This added to the pressure and worsening sense of friction between the two countries (*Japan Times* 1989).

At the same time, however, the number of international companies entering Japan and winning work was increasing, although there was a suspicion that this was being managed rather than left to free competition (*Asahi Evening News* 1989a). By 1992 there were 80 foreign companies with construction licences in Japan, of which 36 were from the USA.

Further complaint from the Americans and the threat of a ban on all Japanese construction companies led to the Japanese adopting an open bidding system for all major projects in 1994, although the USA continued to complain about implementation (*Building and Construction News* 1995).

Comment

As an interested observer it was hard not to be cynical about the objectives of the USA at the time. Despite the nature of the USA complaints there were rumours of the new American firms quickly going native in this respect (*Asahi Evening News* 1989a), while the continual threats from the USA appeared very heavy-handed. With hindsight, however, the net result was, on paper, an open system although, like all things in Japan, practice is often different from the system on paper (see Case Study 5.10).

Cooperation among competitors and clients

It has been reported that the Japanese market has 510 000 construction firms employing roughly 6 million people (Ministry of Construction 1992). All of these firms require at least one of the 28 types of construction licence available. More importantly, it is reported that only 57 of these firms are members of the Japanese Federation of Construction Contractors (JFCC 1993), viewed as the league of major contractors.

In global terms the Japanese contractors have slid from the top five into mid-table positions, the result of a depressed home market during the 1990s. Consultants, by contrast, have made significant strides into global work, since only Nippon Koei and Pacific Consultants had made the global top 50 in 1994 (ENR 1995a). Evidence on building material producers is, as always, elusive (Kuroki 2000).

Table 5.11 represents only a starting point in studying Japanese building material production, since the trading houses are viewed as important players in the market. Extensive research reveals no easily accessed list as we have seen in Europe and the USA. The construction section of the Tokyo Stock Exchange contains a range of companies which include utilities, electricals, steel companies and the above trading companies, all with a mix of products which include construction related goods. A list of the top building material producers (as opposed to traders) is more likely to include companies such as Asahi Glass, Nippon Steel, Kawasaki Steel and others who are difficult to identify.

Table 5.11   Japanese trading houses.

Mitsubishi Corporation
Itochu Corporation
Sumitomo Corporation
Marubeni Corporation
Mitsui and Co.

Table 5.12   Top 10 contractors in Japan.

	Turnover $m	International work %	1999 global rank	1994 rank
Taisei	13,238	4.1	47	3
Obayashi	11,775	14.7	18	4
Shimizu	11,285	8.7	30	1
Kajima	11,190	12.6	21	2
Takenaka	10,117	5.5	46	6
Kumagai	7,676	8.0	44	10
Nishimatsu	5,829	28.2	20	13
Kinden	4,652	5.1	86	17
Penta Ocean	4,471	21.6	32	24
Hazama	4,372	9.3	70	—

Adapted from ENR (1995b) and ENR (1999)

Table 5.13   Top 5 consultants in Japan.

	International turnover $m	% Total	Global rank
JGC	310	69	15
Toyo Engineering	160	78	27
Pacific Consultants	154	26	29
Nippon Koei	141	70	34
Chiyoda	28	56	97

Adapted from ENR (2000a)

Negotiated work represents much of the work in Japan with up to 80% of building and 45% of civil engineering contracts carried out on a negotiated basis (Nahapiet 1995). This adds to the tightly knit relationships and the stability of workloads for the players. In return, the customer focus is legendary.

Fig. 5.7   This wharf at Nakatotsu in Japan was heavily damaged during the Kobe Great Earthquake in 1995. The rehabilitation project involved dredging, piling and the development of this futuristic building. (Photo courtesy of Penta Ocean Construction, Tokyo, Japan)

Design and construction are highly integrated, all under the management of the main contractor. Although it was illegal for major contractors to own design consultancies, there have been clear and close relationships arising from this close working relationship. It is reported, however, that the advent of foreign players moving into the market has accelerated the growth of more independent consultants, similar to in the USA and Europe (Ishii 1991).

The close relationships have led to a number of distinct differences from markets such as the EU and the USA. Apart from attitudes to customers, there is a clear emphasis on innovation, research and development, all more possible in a trusting environment. For example, it was reported that the Ministry of Construction's 162 extra-departmental institutions (Wolferen 1989) contained a large proportion of research interests. The 40 biggest contractors all had their own research institutes.

The closeness of relationships between competitors, however, extended beyond the acceptable, and into the rigging of tenders. It is now widely accepted that a façade of public bidding for large public works projects actually hid a system called *dango*. This required membership of a relevant trade association, with the association then becoming a meeting place for nominated contractors to share out work rather than compete. Bid values and the winning tenderer were decided in advance. The system was outlawed in 1993 (Lenssen 1999; RICE 1994).

It was also widely accepted that politicians, civil servants and the ministries all colluded in this practice. This, in turn, contributed to a system of stable profits and 'pork barrel politics', with politicians openly listing their involvement in the development of projects and civil servants rewarded with jobs (Wolferen 1989). A 1992 survey showed 363 out of 2021 senior managers in the top 61 construction companies to be former civil servants. It is also reported that 5% to 10% of contract values were translated into political donations, resulting in a reported cost inflation of 30% to 50% (Lenssen 1999).

Case Study 5.10: The disappearance of *dango*?

Although officially outlawed in 1993 there has been a strong suspicion that the dango approach has not disappeared. Articles have continued to appear on the subject and clampdowns have continued, resulting, in one case, in fines for 400 firms (*Construction Europe* 1995).

In 2000 a former construction minister was arrested on suspicion of receiving kickbacks in return for public works contracts (*Economist* 2000). The arrest led to the police raiding and searching the offices of the Construction Ministry as the scandal spread.

International market

Foreign players in the market remain a cosseted handful, even though the barriers may have been reduced. The generosity of the Japanese government to its construction industry extended abroad, with the Japanese Overseas Development Agency providing the largest share of development aid in the world (OCAJI 1991), much of it through Japanese contractors. In overseas markets the leading Japanese contractors were winning $6.5bn annually by 1988, 40% of this in the markets of SE Asia (OCAJI 1991). Much of this is in tandem with other Japanese organisations. This is fairly small in scale, representing just over 1% of total turnover for Japanese construction (by comparison the same figure in the UK was roughly 10%). However, the combined effect of large Japanese government aid programmes and the cushion of the home market allowed Japanese construction companies to compete aggressively abroad at the time.

5.2.3   The USA – a case study in joined up regional markets?

The global image of the USA construction industry relies heavily on the extremes of Hollywood movies: either a dirty industry where the lead players are mafia-related, or a principled and scientific pursuit of worthy goals, with the US Army Corps often taking the lead. It is possible that both have small grains of truth in them, although high profile corruption of the sort that regularly leaks out of Japan or parts of Europe is less common.

The nation as a market attracts interest from around the world despite many believing that it is fragmented (Fowler 1997) and highly state focused. The scale and diversity of the USA construction market is important; it was estimated that it involved $810 bn of work in 1999 and that there were 1.9 million active businesses (US Census 2000). National government monitors the industry and provides both national and, equally importantly, state-by-state breakdowns of activity since state government and state scale agencies continue to play an important part as funder and client for many construction projects (ENR 2000b). Thus, the old pattern of major players being strong in one region but not throughout the nation has been replaced by the major USA players now featuring at the top of global lists with widespread global presence.

By contrast to Japan and Europe, which have had much to rebuild in the past half-century and have an increasing amount of work expressed in terms of repair, maintenance or rebuild, the USA appears to still have an emphasis on expansion and new-build to cope with the ever-growing population. The country is so large in geographical terms that this apparently remains a feasible (and cheaper) option. A report from 1991 (RICE 1991) suggests that repair and maintenance represented roughly 22% of the market. By comparison, the EU average appeared to be about 32% at the same time. It is clear, however, that nature can be cruel and rebuild after natural disasters has added significantly to the construction market in recent years.

Case Study 5.11: US market outlooks

The 1992 outlook in ENR (1992) envisaged growth of less than 1% in a weak market, with private construction investment actually declining by 15%, a total market of $380 bn, the strongest regional markets being Midwest and South, and recovery looked for in the mid 1990s.

By 1999 the outlook (US Census 2000) was growth of 6% in a strong growth, total market of $809 bn (figures not adjusted to 1992 values) in a continuing upward trend for the market.

The above mirrors changes in the USA economy over the period. Thus, recession and recovery in the overall economy appear to influence the cyclical nature of construction. RICE (1994) reports that the market rises and falls freely with the economic business cycle, little affected by public sector attempts at manipulation to flatten the cycle.

Another interesting contrast (RICE 1994) between the USA and other markets is the large amount of privately funded work, and hence the importance of private sector clients to US construction companies (Table 5.14). This can create an important difference since the private sector client is often more flexible and amenable to negotiated work and follow-on work. Maintenance of close relationships and a strong focus on delivering projects on time and within budget has become the norm, rather than costly competitive tendering all the time (Fowler 1997). This has an important knock-on effect in international work. For example, Gibb, a UK consultant with a US parent company, report their increased ability to tap into their parent's access to 300 of the top 500 US corporations after the company was acquired by the US group, Law (NCE 2000e).

Table 5.14   Private and public funded work in USA, EU and Japan.

	Private	Public
USA	72.5%	17.5%
EU	60.2%	39.8%
Japan	55.8%	44.2%

Source: RICE 1994

The latest global lists show that 5 of the top 10 international design firms and 3 of the top 10 contractors are US based. In 1994 US firms were less prominent. The emergence of US global players has been through both increased project-based activity and acquisition. It is reported (NCE 2000a), for example, that 6 of the UK's 20 largest consultants have US based parent companies and that American construction design companies win 43% of work won by overseas companies in the Middle East, a key global market.

As well as becoming global names the top US contractors (Table 5.15) show an interesting correlation with the list of US top consultants (Table 5.16). Six of the top consultants feature in the top 20 contractors, an indication of the strength of the multidiscipline groups in the USA and signaling a strong design and build capability.

Table 5.15 US top contractors.

Bechtel Group
Fluor Daniel
Kellogg Brown and Root
Turner Corporation
CENTEX
Skanska
Kiewit Sons Inc
Foster Wheeler
Bovis Lend Lease
Gilbane

Many of the larger players have often started from a strong oil and gas background where the USA has had traditional strengths and there is a scale of activity which allows companies to grow. Whereas Japanese companies have been barred from having both design and construction capability and the Europeans have been slow to develop joint skills, the large US players have jumped into dominating positions in both, with three of the US companies in the global top 10 involved in both design and contracting – Bechtel, Fluor Daniel, and Kellogg, Brown and Root). Thus, an important element of the market may be integrated services.

US building material players (Table 5.17) are less noticeable on the global scene and reference to the biggest 10 is on the basis of market capitalisation rather than turnover. It appears to be an under-researched area.

Table 5.16 US top consultants (numbers in brackets indicate global position).

Bechtel Group (1)
URS
Fluor Daniel (2)
Jacobs Engineering (20)
Foster Wheeler (8)
CH2M
Parsons Brinckerhoff
Parsons Corp (18)
Kellogg, Brown and Root (3)
Earth Tech

Table 5.17 US top building material players.

Nucor Corp
Martin Marietta Materials
Ashland Inc
Northwestern Steel and Wire
Lafarge Corp
Florida Rock Industries
Texas Industries
USG Corporation
Vulcan Materials
Southdown Inc

Source ENR (2000e)

The free market approach to construction

Just as Japan went through a period of being the model of best practice, it appears that the USA market is now the role-model. A number of studies of the market since the late 1990s have reported on the methods of procurement, working practices on site and employment systems as being superior to Japan and Europe. However, even where there is some truth in the ideal market place the findings must be viewed with caution since a study of US based articles reveals similar problems, accidents and failures occurring within the USA as elsewhere.

Case Study 5.12: Boston Artery (NCE 2000f,g,h)

Hailed as the biggest civil engineering project at the time, the main scheme is to replace Boston's elevated Central Artery with a tunnel and associated works below the Charles River in Boston Harbor. The 13 year project had several revisions to its estimate, raising it from the initial $1.6 bn to a predicted cost of over $8 bn. The scale, ground conditions and access problems in a heavily trafficked area have had their effects on the project, and engineering has been complex.

The work was project managed by a joint venture of two of the biggest players in the USA. In total 109 construction contracts were let to complete 260 km of new highway lanes, major tunnels and the widest cable-stayed bridge in the world. One third of the cost was sunk into environmental mitigation and traffic management during the project.

The overspend and spiralling cost have been put down to unforeseen ground conditions in an old heavily used harbour area, measures to placate the local community and the volume of design and change orders, which on average run to 24% extra cost.

A study by the Japanese (RICE 1994) provides a useful overview of the US construction market together with comparisons against Japanese and European practice. Their report suggests a relatively open market, although with some interesting elements which may hinder the entry of non-US players into the market.

It was believed to be openly competitive, although with heavy penalties for failure to deliver. These penalties, which are both incorporated into contractual documents and arise from the more litigious culture of US business, are often stronger than the Japanese or European norm. Interaction and anecdotal evidence suggests that this frightens many away from pursuing entry to the US market. Although there are very few foreign corporations featuring in the top 50 lists for the USA, there are numerous European and Japanese construction companies with a presence in the US market and the barrier is probably more perception than reality.

Case Study 5.13: Odebrecht in the USA (Royse 1998)

Odebrecht, a large Brazilian-based construction company, has made significant strides into both the global and the US markets. It has an annual turnover of $5 bn and has picked up projects across Europe, South America and Africa. It has based its US subsidiary in Miami and has been involved in projects such as the expansion of Miami Airport.

RICE (1994) reports on the heavier perceived penalties which drive delivery in the USA and the strong private sector client base which promotes relationship development. Fowler reports (1997) on the high unit labour costs driving efficiency, mechanisation and standardisation in building services provision. His findings also suggest greater earlier involvement of specialists in the design process and rigorous cost control, although both of these have been cited as weaknesses causing the huge cost overruns at the Boston Artery.

The open competitive culture of the market appears to result in little co-operation among competitors. It has been stated that this is a barrier to innovation in the sector (Fowler 1997), with the USA lagging behind other nations in terms of construction sector research and development. However, investment does occur and the US construction industry has a justifiably proud record of innovation. Much of the innovation is either client led or association led.

Who are the big client organisations? The private sector with 70% of the market is the biggest player. As elsewhere, however, government plays a critical role although there are a variety of funding streams. Key government players appear to be the Department of Energy, Department of Transportation at both federal and its state level equivalents, the US Army Corp of Engineers, the state governments and various quasi-government agencies such as port authorities, airports or bodies such as the Environmental Protection Agency. A key contrast is the de-centralised nature of the system by comparison to, for example, Japan with its very powerful Ministry of Construction.

It is reported (RICE 1991), for example, that in the USA in 1985 the public sector construction spend was split on a $27:23:49$ ratio between federal, state and local. This followed a pattern of increasing delegation to local level since at least 1960. It would indicate, however, the strength of local spend and decisions, rather than the usual emphasis on the role of state funding as a key factor in the market.

Some of the work, as elsewhere, is provided through large-scale super projects but, unlike Japan where they are designed to be leading edge (OCAJI 1991), it has been reported that US projects are specifically designed to be less leading edge (Fowler 1997). The aim is simple, straightforward and proven systems. Given the risks inherent in large projects this is an understandable ideal. Recent examples reported in the international press include the Boston Artery project and numerous airport expansions and upgrades.

A look at articles on the US market (ENR 2000b) reveals that the issues of interest are probably very similar to those found in the European market: squeezed budgets, increasing emphasis on environmental projects, consolidation of the players through merger and acquisition, accidents and problems and the insidious effect of IT on the business.

An international market?

The barriers to entry are more informal than formal and some sectors of the international building material players have made significant inroads into the US market. At one stage UK consultants were very interested in the market although the flow of interest now appears to have moved in the opposite direction with US firms acquiring UK companies.

In overseas markets the leading American contractors win sizeable portions of work globally (Thorn *et al.* 1997). Close partnerships and relationships with both US private sector clients and with the US Government marked their progress (RICE 1994).

5.3   Competitive advantage and thinking ahead

The gist of the message so far has been in learning the tools and spotting the patterns to ensure that a construction business survives without insurmountable risks. However, that is not enough. The common desire in business is to thrive rather than survive, and to do this a construction player needs to highlight its competitive advantage and then make full use of that knowledge. Again this is simple advice given to all types of companies by management specialists, and in that respect international construction businesses are no different.

The difference of course is in the implementation, much of which is discussed in Chapters 6 to 10. At the same time, it is linked to the previous chapters which were essentially about planning and developing a business. The on-going operation and development of the plan is often seen as separate from the start-up phase, since practice intrudes on the theory and all sorts of unknowns come into play which may render the original plan useless.

While I am convinced that flexibility is key to success (and by implication adherence to a rigid plan is unhelpful), there is still an important role for the business plan to play after the initial decisions have been made. It becomes a review document, a benchmark and a possible source of ideas for future directions.

It is probably useful to look at the need for competitive

advantage and then one or two brief examples of such key factors. The reasons for establishing competitive advantage are simple. A business needs:

■ To compete and attract clients
■ To know how to use information to best advantage
■ To identify the best opportunities
■ To know and avoid dwelling on weaknesses

Much of this seems like common sense. However, my own experience is that it is much more difficult to spot these competitive advantages than would first appear. Major companies have been left in positions where they clearly believe that competitive advantage lies in one field when in fact, for various reasons, they have no advantage.

In fact, the approach to competitive advantage can be twofold, which we shall describe as the traditional versus the MBA-type approach. Neither can claim to be exclusively right. There is an element of value in both and it is important therefore to consider both (see Table 5.18).

Table 5.18   Traditional *versus* MBA approaches.

Traditional	MBA-type approach
Early warning systems	Resource basis
Prequalification	Unique selling points
Tender	Systems and visions
Fire-fighting	Customer focus
Project-based	Corporate-based

5.3.1   The traditional approach

The traditional approach is based around the project cycle, the basic building block in construction. There is a belief that a full knowledge of how the construction cycle operates and early knowledge of the progress of particular projects are the most important factors in winning competitive advantage in construction. Projects which match a company's strengths can then be shepherded into the company's basket of opportunity and presumably prequalification will follow. The biggest companies (because they have the resources) and the best (because they have benefited from the system) will often employ an early warning tracking system to track project progress through the cycle.

A key point in the very early stages is the identification of a funding source. This is a useful checkpoint particularly in pursuit of speculative projects since no sign of a funder usually means trouble. More importantly, funding that does not equal cost usually means a difficult gap which often becomes the source of more work than the project itself.

Prequalification

This a critical point since, as we have identified, it exposes a business to questions about its competitive advantages. It is at this point that joint ventures are often developed, primarily to gain qualification rather than for tendering or in planning to complete the work. Prequalification as an exercise is a great leveller. I have read business plans which proclaim competitive advantage in, for example, port construction or tunnelling and then the company has failed to prequalify for major projects in these fields because the in-house expertise has moved on or the most recent experience is too old – fundamental problems which, with hindsight, should have been spotted but still occur.

Tendering

This is the point in the project cycle often viewed as the end-point for converting competitive potential into actual work. There is a strong belief among traditionalists that competitive tendering fosters the best, and competitive advantage is converted into work won. There are others, of course, who believe that tenders are frequently won on omissions and mistakes, and competitive advantage is less important than luck.

Fire-fighting

This would be called project management in any other sector. The simplest source of competitive advantage is having systems and structures which allow the business to deal with problems as they arise. Construction is a business which is built on problem solving and problems are inevitable.

An important key in the traditional equation is remembering the need to cover both corporate and project needs. At the corporate level key issues are security and growth versus the risks and these clearly need to be addressed. At the project level (where competitive advantage is often perceived to be concentrated) there is also a need for security and risk assessment but one project does not sustain a business. Smaller markets, for example, have capability for only one major project in a particular sector every few years, clearly insufficient to sustain the presence of a big international player unless other factors are taken into account.

Corporate needs are fulfilled through a variety of activities and methods. Chapter 4 has already pointed to securitisation and insurances through passing on risk elsewhere. Hedging is another financial instrument used to reduce risk at the corporate level.

5.3.2   The MBA type approach

In the MBA type approach there appears to be a stronger emphasis on the corporate base, the resources and how they can best be employed for maximum effect. The resource which is key, as we explained in Chapter 4, includes finance, asset and human aspects. Emphasis is also placed on developing a culture devoted to growth and looking for growth opportunities. At the corporate level this can extend beyond sourcing projects and into areas such as investment, mergers and acquisitions since these can be methods of buying in competitive advantage. All of these are illustrated through examples in the following chapters.

Increasingly, cultural factors and localisation are viewed as important for the long-term maintenance of competitive advantage and case studies showing this are presented in the following chapters.

The simplest source of competitive advantage is having unique selling points (USP). Again MBA texts wax lyrical about the wide variety of USPs available and there is little point in pursuing these in detail. Corporate competitive advantages which have been well documented in MBA texts include, for example, branding, closeness to client etc. This introduces different perspectives and needs.

USPs which appear to have significant effect in the construction sector and which I will expand on in Chapters 6 to 9 include the following:

Size

Size of company matters in international circles because it adds security and strength in negotiation, contributes to the development of brand (and all the advantages that that entails) and it may lead to economies of scale and market leadership. Again, case studies in the following chapters provide useful explanation of this factor.

Specialism

This is the area which most construction companies seek to emphasise. Many do have specific advantages but, as stated earlier, it is easy to make mistakes through use of old information or the misinterpretation of complicated data.

Forward thinking

This approach, suggested most by MBA texts, has many merits but the jury is still out because of the difficulties of future prediction. Indirectly, evidence of planning suggests evidence of care, which must be beneficial.

Home market cushions/advantages

This appears to be seldom discussed but is a very significant advantage. Closed markets often have cartel implications and can close out technological advance but they can also cushion home construction companies when they expand abroad, as section 5.2.2 showed. The lack of vigorous competition at home implies security of market share (and possibly profit) and provides a cushion which allows them to take greater risk in foreign markets.

Systems and visions

This may be an irreverent manner of lumping together all the MBA expert work on identifying all the internal factors within an organisation which lead to a well-run forward-looking company which succeeds. However, as stated in Chapter 4, it is important to note their effect and their importance where they exist.

Customer focus

This is the holy grail of MBA land. It is an area that many construction companies have devoted much time to in both domestic and international operations. There are many cases of such companies working their way up the customer development process from attracting first time customers through the development of repeat customers and reaching the heights of partnering with clients, on a possible equal footing (Kotler 1997).

To complicate the picture further there are two other examples of competitive advantage which are useful but difficult to support by hard fact. They probably would be classified as falling within the MBA-type approach but the main point is to highlight the difficulties of spotting the competitive advantage.

Case Study 5.14: Vertical integration

The earlier use of Tarmac as an illustrative example (in Chapter 3) established that Tarmac had a vertically integrated set-up, i.e. the company had interests in quarrying which fed into building products which in turn fed into contracting, the expertise from which could be used in consultancy.

Thus, it had an interest throughout the construction value chain, and it was believed that the knowledge developed from this led to an ability to control or influence the cost, demand and choice in the industry, i.e. by having a major interest in the basic material the building product division could be provided with the best priced, best product, while the quarrying division would quickly learn what material the building material division and, by extension, that sector needed. There was, however, a belief that the company used the chain in a much smarter manner. It was believed that the contracting division generated large volumes of cash but very little profit. The cash was then invested in the higher margin building materials division for slightly more profit over slightly longer business cycles before being used again in the highly profitable but cash intensive business of house-building (and quarrying?).

Thus, Tarmac was a recycler of money rather than a construction business. Its competitive advantage over fellow contractors lay in its ability to weather the inevitable cash-flow crises of contracting while, at the other extreme, the generation of cash kept the housing business flowing through quiet periods of house-buying.

Comment

There are other references on this subject which imply that this is a common method of management within contracting circles. However, as the example in Chapter 3 indicated, it is not foolproof. Relationships are another clearly important pillar of success as the next case study shows.

Case Study 5.15: A political deft touch

Bechtel clearly has a strong relationship with the US government (*Building* 1999). The US government has many major interests in world events. Two special interests lead occasionally to opportunity which US companies can then exploit: market opening and the world's policeman. Two recent examples illustrate how the US company can take advantage of this to set itself for opportunity.

The USA strives to keep world markets as free as possible and has actively pursued the matter with other countries such as Japan (see Case Study 5.9). After heavy pressure at a government to government level by the USA in the 1980s, the stage was set for international companies to benefit from Japan's limited market opening in the construction sector. Bechtel was, in fact, one of the first US companies to exploit this and move into Japan (*Asahi Evening News* 1989a). In the 1990s the Gulf War was an event where the USA played its part in protecting a smaller nation from an aggressive larger one. However, the war left a decimated Kuwait which needed help and assistance from the outside world. Again, Bechtel was there to take advantage of its close relationship with government, placing it in a prime position to win work in rebuilding.

Comment

Thus, a close relationship has helped secure two major opportunities for Bechtel. While neither of the above case studies may involve illegal activity they do point to a complicated string of circumstances which leads to a competitive advantage. In a sense, the two examples above are difficult to map because they, like mergers and acquisitions, arise from the corporate end of a business (common to all industries), while many in the construction industry tend to concentrate on the project level.

In conclusion, it is difficult to pin down competitive advantage and to establish the direct link between the patterns of trade in construction and the unique resources of a particular company. However, management consultants will continue to try and the more obvious examples will be illustrated in the following chapters. The traditional versus the new MBA-type approach will continue to generate interest. Inevitably, however, sourcing the projects in whatever manner is available is viewed as the main objective of developing the business – a simple goal.

Problem solving exercises

(1) Find an article about a company which is based in a developing country but which works internationally. What appears to be their competitive advantage?

(2) Who are the biggest players in the three world markets now and what is their market share? How profitable are they? (Hint: see their websites for annual reports.)

(3) Choose a country in South America. Who are the key players?

6 Building Materials and Construction Equipment

6.1 Introduction

Having considered the analytical tools which can be used, the information available and the systems required for a firm to work in an international construction context, we now consider the ways in which firms actually behave in practice. As stated earlier, the construction industry is often viewed in terms of four sectors: the building material producers, the consultants, the contractors and the plant manufacturers. Each has different players and characteristics.

This chapter will concentrate specifically on the markets and players involved in the two sectors which are also manufacturing industries: construction plant and building material production. They are both more industrialised and more global than contracting or possibly even consulting. Case studies on these sectors do appear in mainstream management texts, unlike contracting and consultancy. The first part concentrates on the plant manufacturers.

6.2 Construction plant

Table 6.1 gives a snapshot of the global top 10 equipment manufacturers for 1994 and 1998 which reveals some similar patterns to those which occur in other construction sectors over the same period. The Japanese were replaced by US firms at the top, while European firms held firm over the period, in a very similar pattern to the contracting tables at this level. There are no players beyond these three regions. The sector itself over the recent past has seen many changing patterns of behaviour. After the Japanese expanded into the global market in the 1980s, there was a period of recession for many of the players in the early 1990s before a return to a healthy market in the mid 1990s. In 1998 the South East Asian crisis hit home, affecting major construction plant players as it had other sectors.

Articles on this sector stress the global nature of the business, the

Table 6.1   Global top ten equipment manufacturers.

1994(1)			1998(2)		
Caterpillar	(US)	$14.3 bn	Caterpillar	(US)	$18 bn
Komatsu	(Jpn)	$7.1 bn	Ingersoll-Rand	(US)	$6.75 bn
Liebherr	(Germ)	$1.8 bn	Case	(US)	$5.4 bn
Deere	(US)	$1.6 bn	Komatsu	(Jpn)	$5.25 bn
Volvo	(Swdn)	$1.6 bn	Atlas Copco	(Swdn)	$3.75 bn
Case	(US)	$1.3 bn	Liebherr	(Germ)	$3 bn
Kobelco	(Jpn)	$1.3 bn	Volvo	(Swdn)	$3 bn
Hitachi	(Jpn)	$1.3 bn	Hitachi	(Jpn)	$2.25 bn
O&K	(Germ)	$1.0 bn	Svedala	(Fin)	$2.25 bn
JCB	(UK)	$0.9 bn	JCB	(UK)	$0.75 bn

Source: Wolf 1995 (1)   Source: Thompson 1998a (2)

need to be aware of competitors and new products and the consolidation through merger and partnership which is occurring across the sector; in short a model of MBA behaviour! Their position as major players in associated industries such as mining and forestry and their obvious close association with the automobile sector and its suppliers make them aware of the world beyond construction and able to talk in terms of global management.

Plant manufacturers are becoming aware of a number of issues affecting the whole industry which will particularly impact their sector: increasing interest in mini-plant, chronic shortage of skilled operatives in many locations, environmental pressures, the computerisation of plant control and the increasing move towards rental rather than bought-in plant. All of these demand change and innovation from the manufacturers.

Articles on the sector carry an interesting message about these companies' relationship with the other construction players. Plant manufacturers clearly want to work more closely with the major contractors but the desire to innovate does not appear to be matched by contractors and others. Thus, the relationship seems to be at arm's length at present.

To further illustrate the sector two case studies (6.1 and 6.2) are presented on two of the major players, their current position and future plans.

Case Study 6.1: Caterpillar (Thompson 1998a and Kotler 1997)

The world's biggest player in construction equipment manufacture was formed in 1925 and, by 1998, had 64 000 employees in 197 countries. The 1990s were a period of modernisation and restructuring as the company dealt with the aggressive expansion of Japanese players into many of their markets. Restructuring led to flatter management systems and a stated belief in the importance of training, reward structures and the development of their human resource. The development of non-traditional areas such as financial services acknowledged the major contractors' and clients' desire to offload the risk of buying equipment and replace it with rental, leasing or other forms of shared responsibility.

The business cycle is an important aspect of future planning with an acknowledgement that the business is cyclical and that downturns need planning and action. In the late 1990s the company's research and development spend was about $750 m per year.

A ten year plan envisaged rapid expansion for the first decade of the twenty-first century and development of new global markets particularly in China (with 20% of the global population and, presumably, a demand to fulfil) and the former Soviet Union (with 40% of the world's known resources and, presumably, a supply to exploit). The expansion will be based around joint venture, innovation embracing information technology in both equipment and management systems and the key, the development of strong distribution systems in their new markets.

Comment

This very short potted history illustrates a global company in which there is a clear message of expansion, innovation and risk management. The company is very focused on supply, through innovation, and demand, through the development of a portfolio of national markets at a global level. This contrasts with articles on contractors and consultants which tend to home in on some aspect of detailed engineering.

This company is, of course, the biggest in this sector and may not be representative of the sector as a whole. However, the following case study on JCB, the smallest of the big ten, shows a similar pattern.

Case Study 6.2: JCB (Thompson 1998b and JCB 2000)

JCB was set up in 1945. It is a family-run business based in the UK but has become a global operation with a well-known brand-name. As a result it exports almost 80% of its products. Innovation is a high priority within the market and it works with universities and market researchers to keep the edge.

The company has a wide spread of markets across Europe and North America and it was strong in South East Asia before the crisis of 1998. However, it has focused its future market expansion plans on India (with 20% of the world's population) and South America (with significant natural resources to exploit). It is cautiously entering China through development of a dealership rather than through joint venture.

Safety and environmental concerns have been two strong drivers in the requirements of new products. JCB, like Caterpillar, has seen a big move from contractors towards rental and leasing rather than purchase and this has altered the profile of machine used and brought financial requirements and changes to their operation.

Comment

It is interesting to note that the main focus for expansion uses similar reasoning to Caterpillar but results in a focus on different markets. However, once again, the case study reads like an MBA text: innovation, a global market, close attention to market research and a desire to move closer to customers in development of new products. The spread of business also acts as a buffer to economic downturn as the company manages its business in the grand tradition of portfolio style.

Despite both companies being in the top ten there is a massive difference in the scale of the two companies. The fundamental approach, however, appears to show little difference. It may be that smaller, more specialist plant manufacturers do show a different approach; anecdotally there is a belief that survival depends on working very closely with contractors and providing a complete package of equipment and expertise, as many groundwork specialists do. However, it is clear that the sector as a whole is dynamic, global and well-versed in the language of modern management.

6.3   Building material producers

The main building material players in Europe, the USA and Japan, many of whom represent the biggest global players, have already been listed in Chapter 5. Their products are diverse and the great debate on inclusivity has been noted: the difficulties involved in tracking the steel, concrete, wood or other material which may or may not be used in some form of construction. Thus, the lists often omit important sectors. One survey in the UK of the leading exported products included plastic sheets, paints and varnish, structural steel, aggregates, reinforcement bars, lamps and fittings (EAG 1998). Thus, the sector is a broad church of products, approaches and influences.

To start this review the PARTS checklist (section 3.2.3) is again used, starting with the players. In most developed countries this would show a small number of very big players and a multitude of smaller players. Many of the bigger players will export. The smaller players are less likely to be exporters for many of the reasons outlined in Chapter 2, although there will be a few who are significant exporters. They are often good examples of having specialist expertise as a competitive advantage.

Many of the same items appear on both import and export lists for many countries and this raises fundamental questions of what is the added value of these goods to their markets. It is difficult to state whether this is a failure of market efficiency or a desire for different types of the same product in different countries.

Case Study 6.3: German–UK trade

Looking, for example, at the trade between Germany and the UK the biggest export from the UK to Germany in 1995 was steel products which represented 21% of the total (Mawhinney 1997a). The top five products, apart from steel, were aluminium products, air purifying equipment, electrical building components and paints and varnishes, which represented a combined 37% of the total. Recommendations for the market highlighted the office refurbishment sector, the DIY sector and the low-cost prefabricated housing sector as being the areas of growth although the market itself was going through recession. Thus, for this particular example, goods which were viewed as added value to the German market at that period were concentrated in these sectors, a very cyclical type of market greatly affected by the unification of Germany and the rebuilding of Berlin as the national capital.

The rules of the market are heavily influenced by tradition and standards. Tradition implies the systems of procurement which are prevalent in the market of choice. Are procurement decisions influenced mainly by architects, consultants or the clients themselves? What are the normal trading terms and conditions? Are agents a recognised part of the business and what is their chief role? What after-sales support is required?

The standards applied can be both formal and informal. The case studies on Japan and Portugal in the next section (Case Studies 6.5 and 6.6) provide interesting examples. Formal implies the systems such as British Standards, the German DIN equivalent or whichever code of practice can be applied and what other building standards apply. The informal would rely on how rigorously these standards are applied and how, in general, quality is perceived. This is an important area since inconsistent or uneconomic application of rules can prevent new players from entering a market, a charge made against the Japanese in the 1980s for their prescriptive codes of practice preventing new products from entering the market.

Tactics

Tactics are illustrated through all the case studies in this chapter. The basic approaches available to building material producers are summarized below.

Export

The basic approach of keeping operations in the home country but seeking export markets is still the main thrust for many manufacturers. In general the choice of market should be on the basis of careful planning as outlined in Chapters 3 and 4, although frequently manufacturers seize an opportunity or, even more fortuitously, are invited to sell products in a new country.

Planned approaches would normally research the general economic situation of a country, its construction industry growth, import tariffs and who else is selling in the market. All approaches, planned or otherwise, need research into the specific demand for a product, the prevailing standards of quality, the decision-makers and accepted procedures for procurement, the type of distribution that exists for imported goods, the need for agents, typical payment arrangements and the cost of marketing.

The actual market entry would fall into one of four categories: setting up a marketing office from where 'selling', administration and after-sales would be conducted, setting up a joint venture with a local player where operational support would be provided, employing an agent to seek work or opportunity or employing or working with a distributor who will sell the product in a retail or wholesale outlet. The choice depends on a variety of factors such as the type of market, the need for an after-sales service, the availability of agents, partners or distributors and the corporate desire to control the operation or minimise costs (see Table 6.2).

Project or product approach

This is a similar approach to that employed by many contractors and consultants whereby specific project or product demand is monitored and followed. Examples of this are strongest in sectors

Table 6.2 Export strategies.

Method	Cost	Control
Setting up an office		
Setting up a joint venture		
Employing an agent		
Employing a distributor		

such as the power or even the military sectors where specialist producers of, for example, power station boiler plant or specialist hardened concrete bunkers hold a niche market dependent on a project-by-project approach.

Corporate approach

The approach which is most widely applied beyond construction is to look for a corporate solution through merger, acquisition or other common business method. An interesting example of this has been the cement industry, where all of the top players have acted in similar ways.

Case Study 6.4: Cement industry in Asia (*Economist* 1999; Davis Langdon and Everest 2000; *International Construction* 2000)

The global cement industry is dominated by six multinational companies: Switzerland's Holderbank, France's Lafarge, Mexico's Cemex, Britain's Blue Circle, Germany's Heidelberger Zement and Italy's Italcementi. Having dominated European markets they captured much of the North American market in the 1980s before dividing up Eastern Europe in the early 1990s and Latin America in the mid-1990s.

The South East Asian boom saw the economies of countries such as Malaysia, Thailand, the Philippines and Indonesia develop large construction markets. These markets required a large expansion in cement production to fuel the construction boom. Between 1992 and 1997 Thailand built nearly three times Britain's total cement capacity despite having a market only 30% the size of the UK construction market. Much of this was developed by local companies.

The crash of 1997–8 in the region brought construction to a standstill and highlighted the overcapacity (estimated at 25 m tonnes – one tenth of the production of Europe). Despite the supply-demand imbalance the corporate approach to cement has seen the big six go on the merger and acquisition (M&A) trail across South East Asia. M&A activity was estimated at $1 bn in 1998.

There are a number of drivers in this trade besides exploiting a timely opportunity. The fastest growing markets are the developing countries. The big multinationals reap economies of scale on technology, capital cost and marketing. Although land transport is prohibitively expensive, transport by sea is a profitable business with differences between Thailand ($15/tonne) and West Coast USA ($70/tonne) providing new markets.

Not all of Asia has been conquered although the big six now control an estimated 60% of total capacity. Significant local players still exist in, for example, Indonesia. South Korea has distorting transport subsidies which favour local competition, Chinese players indulge in product substitution and Taiwan remains expensive in M&A terms.

One example of the global giants is Blue Circle, a £2.3 bn global player, who had spent £425 m on acquisitions by December 1998 (King 1998). Purchases included a controlling stake in the Malaysian market and a 20% stake in the Philippines. The company itself had been tracking the Asian market for three years but had felt that prices were too high before the opportunity arose through the slump of the Asian market. Blue Circle see their investment as buying at the bottom of the market, since long-term prospects are for huge growth for the region. The company's other current business development interest is in North America, where again they believe prices are unsustainable and a recession is awaited.

Comment

Domination of the European market has provided the big players with a stable base from which to develop global strategies. The domination, however, has often been accompanied by rumours of cartels and monopolies, although evidence is scarce. That said, however, the long-term nature of these companies' planning would be viewed as top-class by many management graduates.

Scope

Scope is very varied as we have already discussed. Players range from the old Tarmac which had a foothold across much of the vertical basic materials chain through to specialist producers of doorknobs. In many ways, the scope of a company's operations provide a clue as to its best approach – small one product manufacturers look at exports while big conglomerates can look at corporate approaches.

6.4 How is the building material sector different?

Developing and operating a manufacturing business has many similarities but also many differences from contracting or consultancy.

6.4.1 Business development

Interfaces with the client are both direct and indirect and in this respect are no different from in other sectors. Two case studies are presented here, based on export advice given by government.

Case Study 6.5: Japan (EAG 1998)

Preparation, patience and after-sales are viewed as key components to exporting into Japan. The Japanese are global procurers but are renowned for putting quality over price, although the recession of the late 1990s has changed that fundamentally. Japan's huge construction market is, through its scale, attractive although it has not grown significantly in recent years.

As part of the opening out of the market, building materials have become an important area of interest, although much of the Japanese interest focuses on niche products rather than bulk purchase. This has included, for example, the product support for the building of traditional British style pubs, or specialist requirements for bricks in a variety of colours and textures. Bulk products, by comparison, still suffer from apparent protectionism.

Comment

There is recognition of the need to take account of varying tastes (sociocultural factors). There is little mention of the main forms of distribution or production employed; typically agents or joint ventures are the most popular method at present. Japan had had a reputation of being a closed market in traditional products and much of the emphasis was on the growing desire of the Japanese to import novel materials. A contrasting example is Portugal, a smaller, less developed nation.

Case Study 6.6: Portugal and its European funding (EAG 1998)

The major item of note in export promotion to Portugal was the £18 bn funds allocated by the EU to structural and infrastructure improvements, contributing to a booming market in construction. It was a rapidly expanding (5% growth 1993–97) though still small market (£7 bn in total). The boom was creating a shortage which, together with EU procurement rules, was creating a fairly open market for exporters from other countries. Neighbouring Spain exported substantial amounts of building materials to Portugal, with the top products being ceramics, sanitary ware and hot water installation equipment as Portugal's boom was holiday and house led.

Comment

European funding and shortages were combining to rapidly open the market for exporters. Again, some sectors within the building materials market were more open to new materials than others, and again advice generally pointed towards appointment of distributors as the first choice.

6.4.2 Operations

Since a large part of the work of the building materials sector is in the production of goods, this sector, like other manufacturers, has considerable fixed costs. This represents a large part of the business and is influential in the choice of market entry.

Variable costs can be important but it is in choice of market entry, influenced by the fixed costs, that differences from the other sectors can emerge. The variable costs are, of course, handled by exercises in efficiency, effectiveness and looking for economies of scale, the trademarks of an MBA approach to establish additional competitive advantage.

Case Study 6.7: Redland – M&A mania (Morby 1997, Davis Langdon and Everest 2000)

In October 1997 a hostile corporate bid was raised for Redland, a major UK building materials group, by Lafarge, a French building materials group. What was the background to this ultimately successful bid?

Redland was a company which rapidly grew through the UK construction boom of the late 1980s. From a portfolio including bricks, roof tiling, aggregates and china clay products it launched itself into plasterboard manufacture in 1988. At around the same time it started to acquire smaller building material producers in France and Germany to build up a European presence. By 1992 it had become the UK's biggest materials group. However, a failing plasterboard business and simultaneous slumps in the UK and Germany brought problems of cash-flow, forcing changes; a pull-out from the plasterboard market (1992), the china clay operations sold-off (1995), a joint venture stake sold in the roof tile businesses (1996) and the brick-making business broken up and sold off (1993–6).

By 1997 Redland was being valued at £1.7 bn (city analysts expected Lafarge to pay 350p per share). Its share price, which had peaked in 1991 at 650p per share, had fallen to 230p per share at the start of the hostile bid in 1997. As a result of this Redland had started a restructuring review and was warning that it was vulnerable to take-over because of its share price. Lafarge were, in 1997, believed to be targeting the UK market for expansion of their operations. It was believed that they were initially seeking aggregates and cement operations, and city analysts believed that Redland was a useful target on the aggregates side.

The aim was to create a group with combined sales of £6.3 bn and a major presence in the USA, Germany, France and the UK (the leading markets in Europe and America). It would benefit from the economies of scale and market domination afforded by being the world's largest aggregates producer, Europe's largest roof tile maker and the world's second largest ready-mixed concrete operations.

In response to the Lafarge bid it was expected that Redland would mount one of two possible defences: the announcement of a restructuring plan which would include further sell-offs and the development of new products, persuading its shareholders to hold on to their shares and not to sell to Lafarge, or it could seek to encourage a friendly bid from a major UK player such as Hanson. The second option, while acceptable in a European context, could be challenged by the UK authorities since any combination of the UK players was likely to develop into a monopoly for a number of product ranges. The bid emerged successful in 1998.

Comment

At this corporate level the building materials sector has had a reputation of being much healthier than the contractors, one of the reasons that Tarmac was put under so much pressure to demerge. In a previous study on Tarmac, which straddled building materials and contracting (Mawhinney 1997b), the difference was highlighted, since contractors showed profit levels of 1% to 3% and building materials were typically at 5% to 10%. Thus, the sector appears more stable in terms of risk and reward.

Problem solving exercises

(1) You work for a company which specialises in waste water pumping equipment. You believe that the company would be interested in the opportunity outlined below in Latvia. If the time-scales were such that you were provided with only a day's notice before a visit to the market to assess the opportunity, develop an overview from the worldwide web of the main in-country risks, threats and other possible opportunity. Describe the preparation needed for the visit – who to talk to, what to talk about, places to visit, supply options, etc.

The project

The municipal authority in Riga, the main town in Latvia, has received funding from Global Development Bank for the design, supply and implementation of the refurbishment of a waste water pumping station. The total average dry weather inflow for the station is 4m³/sec. The tender will call for the provision of three constant and three variable waste water pumps and associated equipment, together with all M&E equipment needed for the works.

(2) Pick a sector such as steel. Who are the main global players? (Hint: see the websites of trade associations in this sector.)

7 Consultant Case Studies

7.1 Introduction to markets and players

We have already looked briefly at the global lists of the biggest consultants, and at the situations in Japan, US and Europe. To summarise, the global players include mainly European and US design consultants, many with strong links to oil and gas contracting companies. The top global players have, in general, a very wide geographical spread, and are seldom dependent on their home markets. The size of annual turnover in 1999 for the bigger players is generally of the order of over $100 m.

To look in more detail at the wider design and consultant sector we rely again on a series of case studies introduced through the PARTS framework (see section 3.2.3). The review of players in the bigger developed markets revealed a number of similar trends with a few big players, many of whom operate in a global context, and a multitude of smaller players (less than 500 employees). Two examples of this are the UK, where 87% of the biggest players would be classified as small businesses (NCE 2000), and Japan in Chapter 5 where the average number of staff per company appears to be 58 (JCCA 2000).

Japan and the UK also represent good examples in terms of global spread of their consultants. In Japan less than 1% of total workload for Japanese consultants is attributed to overseas work. Much of this is related to Japanese Government aid packages. In the UK the top 10 players have over 50% of their workload overseas. Language and tradition possibly play a large part in the relative positions of these two countries in relation to overseas work. The English language and the aftermath of the empire have combined to make UK consultants a worldly bunch, giving them a long-established toe-hold across continents such as Asia and Africa.

It is also useful to do a quick and rough analysis of sales per employee in the two countries (see Table 7.1). The figures in the table can be interpreted in a number of ways, although caution is required. Clearly there is a huge variation in the value of work between UK and Japanese bases. The general interpretation of the

Table 7.1   Sales per employee.

	Domestic	Overseas	
UK	$45–60k	$75–90k	(based on analysis of figures from NCE 2000)
Japan	$190k		(based on figures from JCCA 2000)

UK based figures would normally be on the basis of efficiency, i.e. there is a general trend of more work completed by fewer staff in overseas projects. This would suggest that overseas operations would appear to be higher value added and more efficient although this must be treated with caution without knowing the costs involved.

Although the difference between total and overseas turnover per staff is noticeable (15% to 25% on average) smaller companies show much larger variation. Random examples in the UK would include the Carl Bro Group figures which suggest a total fee per employee of roughly $50 000 and an overseas equivalent which is 330% of that figure, while Steer Davies Gleed report a difference of 65% between the two (NCE 2000).

It is possible that this difference arises from smaller companies' desire to avoid costs by retaining skeleton staff overseas and focusing tightly on flying in specialist expertise from the domestic base (Carl Bro report only 35 staff based overseas and Steer Davies Gleed report 40 staff overseas). If this is the case then variable costs will be the number one priority.

The bigger companies, by contrast, have large localised offices with large numbers of staff working on a variety of types of work. Since fixed costs will be much larger there is more likely to be corresponding concentration on corporate branding, economies of scale, etc.

Indeed, it is the importance of technical expertise characteristics of the consultants which allows small players to compete in international markets, and this, combined with cost structures, marks them out with a fundamental difference from the other sectors.

Case Study 7.1: Yolles (O'Sullivan 2000, Yolles 2000)

An interesting case study is provided by this company, a small Toronto-based structural engineer. Although having only 156 staff it has a hand in international projects valued at over $4 bn. It has clearly worked hard to establish a leading edge-type niche, picking up awards for being a well-managed company and attracting growing press attention.

The company claims its secret lies in its size, forcing it to retain close relationships with clients, and in a dedication to specialising in blending architecturally-driven design with solid structural engineering. Its European operation is run by a woman, at the time of writing a notable distinction from many rivals. The work on each project is split across two main offices and local sites, driven by a concentration on effective use of resources.

Comment

There are a multitude of small companies who have established a niche like this, allowing them to attract work rather than having to seek work. However, there are only a few who, to date, have managed to create a worldwide spread. Language and reputation clearly help.

Added value

Added value often stems from this technical expertise as noted above. However, increasingly, the consultants are recognising the virtues of cost, flexibility and other MBA-type issues, as the following case study of the world's possibly best-known consultancy name shows.

Case Study 7.2: Ove Arup expansion overseas (Parker 2000; Wolton 2000)

By the year 2000 Ove Arup had become a major global player with almost 5000 staff, 40% of whom are based outside of the UK, its home base. The company had seen steady growth of around 5% in overseas work throughout the late 1990s and had built a worldwide reputation for high profile buildings and the technical engineering associated with big infrastructure. There are significant operations across South East Asia, Europe, Russia and the USA.

However, these headlines hide some other more interesting developments. Staff numbers in the UK have remained fairly static in the same period, and specialist skills such as rail engineering and acoustic, fire and environmental engineering have become mainstream parts of the business. While both of these trends appear to indicate a short-term change in direction, the company insists that they result from long-term planning. Ove Arup is a partnership owned exclusively by trustees and employees. The partnership policy is based on no borrowing from banks, with expansion and new development funded entirely through internal resource.

This is fairly conservative from a financial viewpoint but is claimed to promote a long-term view, with research and development-type activity a prominent source of business development opportunity. Recent examples include the development of a re-usable offshore platform or the commercialisation of specialist expertise on spent nuclear waste containers, neither area associated with Arup's current mainstream reputation.

Ove Arup are actively promoting a policy of 'indigenous leadership' – overseas offices run by local rather than purely expatriate staff. Staff with potential are moved around the global business, with some overseas offices building up skill bases which may not even be available within the UK, the current centre of activity. An interesting example of their approach is in Russia, where they have left a relatively young engineer from the UK to develop the company's Moscow office (Oliver 1998).

Comment

The above is typical of any global multinational developing its own staff resource, where the end objective is to become a multicultural organisation with a Board of Directors to reflect this. The cost to the business is in the requirement to set up large offices across the globe, with significant fixed cost. The advantage, of course, is the ability to get close to clients and to make best use of local skills and staff.

The reputation for quality engineering remains core to the business, and is viewed by management specialists as being high cost but high value for the organisation. Its lack of focus on one particular sector can be viewed as both strength and weakness:

critics would point to a lack of focus while supporters would point to a portfolio approach which balances risks.

Rules

The rules of the market are markedly different for consultants, compared to the other players in construction. The basic service provided by many involves intellectual property, which is difficult to protect and contract arrangements often fall well short of full protection. The work relies heavily on creativity, interpretation of rules and communication, all of which can be hampered by formal and informal barriers, or which can be turned to the advantage of opportunistic players.

Case Study 7.3: Japanese consultants at Kansai Airport

In Chapter 5 the Japanese market and its barriers to foreign entrants were studied. Kansai Airport project was used as a case study, where much was made of foreign participation in the terminal project. However, a key player in the terminal consortium was Nikken Sekkei, who brought knowledge of Japanese standards, an ability to communicate with the client and contractors in the Japanese language, and an element of local knowledge and technical expertise (Normile 1992).

Nikken Sekkei (2000) is a large Japanese consultant employing 1950 staff who conduct a wide range of environmental and construction related consultancy services. According to ENR figures (see Chapter 5), it was the largest Japanese consultant in 1994 with a turnover of about $440 m, a major portion of this possibly Kansai Airport. By 1998, however, its turnover had appeared to decline to $200 m although its overseas presence had grown enormously from 29% to 70% of total turnover. It is therefore one of the smaller global players.

Comment

The rules for working in Japan, both formal and informal, undoubtedly created a role for a Japanese consultant in this prestige project. Again, language, codes and knowledge of the players all played their part. However, having been presented with the opportunity of forming a key part of such a large international development, the consultant appears to have created great value from it, in follow-up action, by developing its international profile.

Tactics

Tactics are illustrated through the case studies which follow. Again, subtle differences from the other sectors are sought, since the emphasis is on service rather than manufacturing or service and manufacturing.

In business development, the interfaces with the client are both direct and indirect, and in this respect no different from other sectors. However, the consultant has always had an advantage in having better access to clients because of the nature of his work (often translating the client's ideas to a design) and therefore people skills are or should be important. Technical experts may not always recognise this, but certainly the big players have come to realise the value of this.

Case Study 7.4: The first truly global player? (Winney 1998; Maunsell 2000)

Since its launch in the UK in the 1950s Maunsell has always had international ambitions. Its workload has been heavily weighted in favour of overseas work for many years, built primarily on transport projects funded by aid banks, public sector and now, increasingly, the private sector. They have a very active 'globalisation' policy, actively employing and promoting local staff. The company has seen itself as a global rather than British company for a long time (at the time of writing only 2% of its fee income comes from the UK (NCE 2000)), and it has its global headquarters in Hong Kong, although it is registered in the Channel Islands.

Offices globally are organised on a skills rather than location basis; for example, Birmingham, UK, is the centre of roads main-tenance and Bangkok is based on metro experience, although expertise is quickly moved across the globe as, for example, when Maunsell won work on the Copenhagen metro.

Bridges are a core expertise with major ongoing work in 1998 including Greece, the Netherlands, Hong Kong and China. The strong link with Hong Kong has been exploited to develop a market in China, and again the company see specialist engineering skills as being key to development in that market.

Comment

It is difficult not to form the opinion that Maunsell are a company built on basic specialist skills: an old-fashioned engineering con-sultancy run by engineers. Articles on the company invariably concentrate on technical skills or projects. Winney (1998) quoted above noted that, despite high-technology IT design resources throughout the company, its new finance director in 1998 found an accounts section with 30 staff and one PC.

As a company, however, it delivered 18% growth and 6.3% return to its shareholders in that year so the overemphasis on engineering did not appear to hinder it unduly. On a business development note my own experience would suggest that many construction clients welcome technical experts more than corpo-rate business types. Others have developed more corporatist approaches where technical expertise is a necessary but secondary consideration.

In terms of operational considerations, since a large part of the work of this sector is in the production of intellectual property the sector is a service industry first and foremost and plays to those rules. Fixed costs, although significant at times, are clearly less important than, for example, for the building material producers. This point can be influential in the choice of market entry inter-nationally since it greatly increases flexibility.

Variable costs are therefore important and often determine choice of market entry and differences from the other sectors. The variable costs are, of course, handled by exercises in efficiency and effectiveness, the trademarks of an MBA approach to establish additional competitive advantage.

Increased flexibility has resulted from IT and has improved consultants' ability to send information across the globe. It is increasingly becoming common practice for consultants in the UK to have the work completed in an Indian office, returned to the UK and passed on to a client in, for example, Indonesia, all through IT. W.S. Atkins, Maunsells and other big UK players can provide examples of this type of approach (NCE 2000).

Staff costs are therefore a critical factor. It is interesting to note that despite continual complaint in the UK about low salary levels in construction, and particularly in consultancy, the wage levels offer significant advantage to UK companies working in international markets. International comparisons often show UK wage levels in the bottom half of developed countries, low enough to compete with competitors but still high enough to need to seek local staff in developing countries where possible.

The question of risk *versus* reward is a difficult one. The nature of consultants, either hidden within group accounts, private companies with no need for publicly available accounts or companies too small to have strict reporting requirements, has made it difficult to extract information on their financial performance. The few references available suggest that the sector produces a wide variation of results, although generally similar to contractors in the 2% to 5% net profit range. This would suggest that risk is significant.

Scope

The scope of work for consultants is clearly very varied but continuing to grow. A case study in Chapter 10, the Jamuna Bridge project (Case Study 10.1), highlights the long gestation period for multilateral aid projects. Consultants can and do benefit from early participation in these projects, winning follow-on work. However, the price for this is the development of a wide set of skills held in reserve, should the project proceed.

As noted earlier, players range from the likes of Ove Arup, a multidisciplinary practice which could rightfully claim expertise in most construction-related subjects, through to quantity surveyors, specialist project managers or specialist consultants. The lessons are that the big seem to practise portfolio management, as Ove Arup show, while the small practices possibly move towards greater specialism. As a sector, however, it is too diverse to generalise.

Most importantly, the case studies show that scope, particularly for the bigger players, is rapidly changing as it is for the contractors (to be studied in the next chapter). The scope for a bigger player

can now encompass a much broader field of activity, from inves-
tor/developer roles in private finance, through management of
facilities management and minor maintenance. All this, of course,
adds new risks and responsibilities.

Before leaving the consultants it is useful to look at the only
global top 20 company which does not come from a developed
nation base – Dar Al-Handasah (2000).

Case Study 7.5: Dar Al-Handasah

From conception in Beirut the Cairo-based company Dar Al-
Handasah has become a major global player, having sat in the top
20 rankings throughout the 1990s. Ranked the biggest in the
important Middle East market it also has a strong presence in
Africa (ranked third), the UK and the USA (where it is ranked
fifth). Its strong origins in the Middle East, where connections are
important, have allowed it to win numerous prestige projects such
as Beirut International Airport. However, it has expanded into the
developed world through, for example, the acquisition of well-
known US companies, making it a major force in that market.

The company itself sees its strengths in terms of the 2000 multilingual staff, a unique cultural perspective (important in the
Middle East) and technical innovation.

Comment

Although technical innovation is claimed by all consultants, cultural perspective and multilingual abilities are undoubtedly qua-
lities which mark this company out from other previous case
studies. However, the more important aspect is their growth into
developed markets, in contrast to the established patterns highlighted in Chapter 5. Is this the way of the future? Cost bases,
developments with IT and acquisitions of foothold companies may
well allow stronger developing nation companies to pursue a
similar pattern. Already, it is acknowledged that the Indian market
may well become, through IT, the back-room office for many major
consultants (Macneil 1998).

Problem solving exercises

(1) You work for a small UK consultancy which operates world-wide and specialises in waste water disposal systems. A development bank funded project (see below) has been advertised for a project in Serbia. It is initially believed that the work will entail 4000 manhours of design, a geotechnical survey, one engineer in Serbia for 10 months and a director visiting for 5 days a month.

The project

The Serb National Bank has obtained funding from the Global Development Bank for the improvement and development of water supply and waste water disposal systems in seven cities in Northern Serbia. The proposed projects will require provision of equipment, construction and implementation of works and services which will include:

- Sewerage collector mains for the seven cities
- Pumping stations
- One waste water treatment plant for 98 500 person equivalent
- One submarine outfall

It is to be noted that the area is one of immense archaeological and tourist value. A consultant is now required to carry out the following services:

- To research the basic data to develop the final design
- To prepare the technical documentation for location approval
- After approval, to prepare the final design and tender documents
- All of the above in accordance with Yugoslav legislation and regulations

(a) Using the table below prepare an initial estimate of possible costs. If word is received that an Italian-led team will bid for the work at around $80 000 how will this affect your bid?

	UK	Italy	Germany	Serbia
Engineer annual salary	$29,500	$18,800	$40,600	$2,900
Director annual salary	$37,000	£31,900	$82,050	$11,400

Assume 2500 hours per year but expatriate status is equivalent to twice usual cost

Assume site survey will cost $10 000 in total

(b) Prepare a risk assessment and evaluate the risks involved in preparing an estimate and tender for this project. What further information is required for a full assessment?

(2) Your company, based in Europe, has been awarded an emergency consultancy contract for a design and supervision project in Pakistan, as outlined below. It is your first venture in the Indian sub-continent and the contract has been awarded based on your past ability to mobilise staff and resources quickly for other international projects of a similar nature. Describe the first steps you would take to prepare to start the design work. You will need to take account of the general information needed, the options available and whether partners are required. Assume that there is only limited knowledge of the site at present.

The project

Global Development Bank (GDB): Emergency Transport Reconstruction
Project name: Karachi Emergency Transport Reconstruction Project
Contact: A.Iqbal c/o GDB head office
Description: The requirement is for a firm of aviation sector professionals to assist in all aspects of the rehabilitation of Karachi Airport (following a riot and subsequent fire damage). The work will include civil, mechanical and electrical works procurement, implementation and the re-establishment of services.
Funding source and status: $1 m funding provided by GDB for the provision of consultancy services. Consultant has been appointed.

8 Contractor Case Studies

8.1 Introduction to markets and players

Contractors are often viewed as the core of the industry, and we have already studied their approach to operations in Chapter 4, reviewed some case studies in Chapter 5 and commented on the listings of major contractors in the UK, Europe and globally.

In this chapter the approach will be to concentrate on the companies themselves and how they have influenced and been influenced by the international market. Despite recent changes, the trend with contractors and contracting markets has often been more localised than in the other sectors, and this still rings true across the globe. Table 1.3 showing the major global contractors in the mid 1990s, for example, shows Japanese contractors as dominant despite their small percentages of overseas work. Thus, 'localised' applies even to the so-called global.

Why is there such a difference between the contractors and others? The PARTS checklist (section 3.2.3) is again employed for an explanation, starting with the players. Most national markets show a number of very big players who operate nationally and possibly internationally, and a multitude of smaller players, in a similar manner to consultants and building material producers. However, unlike these two sectors, the smaller contractors appear very unlikely to be exporters, as we have discussed before. Site operations, upon which all depend, are local in nature and clearly lend themselves to those with local advantages. Only well thought-through economies of scale or clear specialist strategies can help to offset this advantage.

The UK market and its creeping internationalisation is a useful example of key international players exploiting an opening market. As Case Study 5.5 on the Channel Tunnel Rail Link shows, the market has changed although the cause appears to have been pressure from outside rather than any internal desire, the opposite of consultants and producers.

Case Study 8.1: The invasion of the UK (Billingham 1996; Barrie 1998; Whitelaw 2000)

It has already been suggested that the contractor field is still viewed as the least international of the three UK construction sectors. Major European and US companies have therefore viewed it as ripe for exploitation. Two of the most aggressive players initially were Bouyges from France and Bilfinger + Berger (B+B) from Germany.

In 1996 a Bouyges subsidiary started work on its first UK contract. The company's aim was to be a UK top 10 contractor within three years through growth rather than acquisition. By 1998 the company was bidding for and winning PFI (Private Finance Initiative) work with a staff of 60 based in London. With a balance sheet of $15 bn the company was much stronger than any of the UK players, an advantage in the PFI field. At the same time it was looking at joint venture possibilities, aware of the need to capture local knowledge.

B+B arrived in 1990 and gradually assembled a UK contracting business. Its original intention had been to land a headline project but when this failed a strategy of gradual evolution was developed. By 1996 it had built up a turnover of $45 m with 120 staff. Its first contract was for a German client, a convenient way of overcoming the prequalification problem of having no track record in the UK.

By 2000 neither Bouyges nor B+B had made it into the top 50 rankings (NCE 2000). However, in the late 1990s major US contractors were invited into the market by leading clients. Their remit was to trouble-shoot on particularly complex, troublesome rail projects. The emphasis has been on integrated planning teams and an 80:20 rule of 80% tried and tested technology and 20% new.

Comment

The original driver for continental firms' interest in the UK, and UK firms' interest in Europe, was the advent of the single European market in 1992. In practice, however, political resistance to open invitation to non-local players, difficulties in seeking prequalification and differences in construction methods all contributed to making the transition to new markets very difficult and the failure rate was high.

The Americans, by contrast, were invited into the market by high profile clients eager to try new ways and they have formed close relationships with the clients and concentrated on tried and tested technology, typical of a US approach as noted in Chapter 5 (Whitelaw 2000). Early indications are that the approach has been successful. It is interesting to note that one of the main US run projects has, in fact, only 7 US nationals out of a total of 80 managers, lending weight to the theory that local knowledge is often very helpful, even if systems are imported. Ove Arup was put forward in the last chapter as the best known name in consultancy, and it is interesting that the above case study involves three of the best known names in contracting – Bechtel, Bouyges and Skanska – all of whom have moved into the UK market in recent years employing very different methods. Bechtel used client relationships, Bouyges emphasised specialist expertise and Skanska has acquired local companies. All three will be studied in a little more detail in this chapter.

Added value

Added value is a particularly challenging issue in this sector. Contractors offer neither a manufactured product nor a clear service. Often their offer is management expertise or an ability to combine manufacture and service. Thus, the analysis of competitive advantages raises much difficulty and the UK, in recent years, has been characterised by much navel-gazing by contractors on what their offer entails. This debate has been exacerbated because consultants and material producers have clear explanations of their services, and this has led to extra pressure on companies such as the old Tarmac which had a presence in all three sectors.

The US and European invasion of the UK highlighted particular areas for exploitation; for the Americans there was a chance to build on their reputation for project management while the Europeans had scale and the opening of the market on their side.

Case Study 8.2: Skanska – the first truly global contractor?

Skanska has an impressive spread across the globe which warrants attention. It sits in the top 10 in both European and US rankings, a unique position. The favoured starting point for many contractors, a large, well-padded domestic market, is not a comfort to which Skanska has access. Its home market was estimated at $25 bn in 1994 (Davis Langdon and Everest 1995). Even allowing for inflation this is a very small market for a company that by 1999 had turnover of $9.25 bn (Skanska 2000); 72% of its turnover is overseas (50% in the USA).

A search through its website for information on strengths reveals few self-congratulatory statements, with the only clues being numerous articles in support of financial rigour, innovation and co-operation.

Comment

This very successful company provides few clues to its performance beyond the impressive market research displayed on its website, which may signal a rigorous MBA-type management approach and strong IT basics. It appears to concentrate on overseas growth in developed markets where risks may be smaller rather than the usual targets of developing markets. However, it does have a presence in developing markets. The home market provides few benefits beyond a reputation for good technology arising from the requirements of a harsh climate (Davis Langdon and Everest 1995).

By contrast, Bouyges has the background of a strong domestic market. It is therefore useful to look at this company as a case study.

Case Study 8.3: The French giant

The biggest European contractor, Bouyges, has grown quickly in scale, geographically and in diversity. We have already examined its approach to the UK in this chapter and noted in an earlier chapter that the French market has been the subject of cartel enquiries (Case Study 5.7 in Chapter 5). Founded in 1952 the company now has 100 000 employees and a turnover in 1998 of roughly $15 bn. Although a public company, the majority shareholders are the Bouyges family (PSIRU 2000). Nearly half of its construction activity is overseas.

A search through its website (Bouygues 2000) establishes that the company is a conglomerate, encompassing a wide range of activities. The strengths are stated as management of major projects and being able to offer a fully rounded service. Geographically, the old French colonies feature significantly as bases.

Comment

Bouygues appears to be a very traditional general contractor moving with the times. The emphasis in the UK, for example, has been on PFI, a major growth market requiring big contractors with strong finances. However, the company also has within its group well known brand names in the more traditional ends of the construction sector.

Rules

The rules for contractors are usually core to the whole market, and as such they have been identified and explained in Chapter 4. In this chapter the emphasis is therefore on the practical implications on the ground. Two case studies are presented which illustrate the effects of differing rules on international work. The first highlights the cultural differences brought about by different rules and the second looks at the rules in the background.

Case Study 8.4: Indian construction sites (Cook 1998)

In 1998 India was viewed as a great prospective market by European construction companies. Alfred McAlpine was one of the pioneer companies and by 1998 was working on its second five-star hotel in joint venture with a local hotel company. Their approach to the project was a familiar one for western companies operating overseas: their primary role was in a construction management capacity.

Five-star hotels require fast, high quality construction. In developed countries advanced materials and techniques would be the norm. In India, however, labour is very cheap and female. In 1998 the wage rate was $1.50 per day. Skilled labour, which is typically male, is $3 per day. Labour rates dictate labour intensive methods, but this reduces building times.

McAlpine has dispensed with the traditional method of architect as project manager, and their site management controls movement onto site and access rights for all trades, all normal for the UK but new to India. They have replaced traditional bamboo scaffolding with Western style steel scaffold, and introduced greater mechanisation than would be normal in India, all of which has added to cost but reduced construction on this project from five years to two years.

Safety is a major difference from Western standards. Hard hats need flat tops so that labourers can carry concrete on their heads, a traditional way. The usual practice of working barefoot has been banned, although preference is for flip-flops rather than boots (Cook 1998).

Up until 1999 profits earned in the local currency, rupees, could not be converted into other currencies. It was therefore impossible for outside contractors to repatriate profits into their home currency.

Comment

The most significant difference for the Western audience of this is the contrasting attitudes to health and safety. Other notable differences include wage levels which in turn highlight contrasting attitudes to mechanisation and this, in turn, highlights differences in the balance between cost *versus* time *versus* quality. Finally the rule forbidding conversion of profit is significant for many companies who rely on shareholder support; shareholders suffering cuts in dividends are not happy!

The next case study has become famous because it has indirectly contributed to changes in agreements on aid and trade provision (see Chapter 9). The project was part funded through an aid and trade package and appeared to be linked to an arms deal between Malaysia, the host and the UK, whose contractors were lead partners in the project.

Case Study 8.5: Pergau Dam (Jones 1997)

The Pergau Dam has become a historic and controversial project, sitting on the boundaries of acceptability in the provision of aid by a developed nation to a less developed nation. Completed in 1997 the hydroelectric scheme has a 750 m high dam creating a 53.3 m cubic metre reservoir, and it cost £414 m. Australian consultant, Snowy Mountain Engineering Corporation, developed the tender design in 1993 although the actual work was completed as a design and build project by international contractors Balfour Beatty and Cementation (part of Kvaerner) in joint venture with local contractor Kerjaya. At its height the project employed a multiracial workforce of 4000.

Technically the project is not large although logistics were an obvious problem since the site is located deep in jungle. It was decided to set up camp for the workers at a remote part of the works. There have been adverse effects although the local population were reported to be happy with the completed project because it had brought improvement to local infrastructure and the local economy. However, the high wage levels cannot be matched by other work in the area and so local workers have moved on and the access roads have opened up the jungle to loggers and developers.

Comment

Although an impressive engineering feat the project became notorious because of the off-site project financing arrangement. It appeared that despite studies showing the scheme was uneconomic the project went ahead, through support from aid and trade provision. The result was an eventual change to tighten up the rules on aid (see Chapter 9).

Tactics

Tactics available for contractors to succeed in the market have been discussed in Chapters 4 and 5. From specialist expertise through to use of corporate muscle, the sector shows a wide variety of approaches. Many continue to rely on seeking opportunity and keeping costs down in a basic approach characterised by contracts, contacts and turnover. With turnover comes cash-flow which appears more important than profit in a low margin sector. A few, however, exhibit different approaches. Two extremes of how the smallest and the largest seek new markets are therefore examined in the following case studies.

Case Study 8.6: Exporting to the Caribbean (Parker 1998)

An example of a small company approach to overseas work is John Martin Construction (JMC). This was a small UK regionally-based general civil contractor set up in 1967. By 1987 it had reached an annual turnover of £3 m. In that year, one of their projects was a quay restoration in Harwich which required some difficult temporary works in the repair of quayside sheet piling. The firm's engineer came up with the idea of a re-usable limpet dam, ideal for quayside work which had previously relied on divers, tides and access. The technology was developed in-house and rapidly became known across the UK.

Thus, when a similar problem arose in Grenada, the company was invited to look at the repair alternatives. The result was a £900 000 ($1.4 m) contract. JMC loaded up all specialist equipment necessary into eight containers in the UK, including two specially built limpet dams. In Grenada they worked with a main contractor, who provided a crane and generators to get the dams into place. JMC then completed the specialist work within the dam.

Comment

This was an excellent example of a project-based, small-scale opportunity which was realised. The anecdotal 'all in a container' type story is often just a famous myth, unless it relies on specialist added value, more in keeping with consultants' approaches than big contractors, a strength that even the minnows can exploit.

At the other extreme is the world's largest contractor, another useful case study.

Case Study 8.7: The world number one

An explanation of international construction would not be complete without further reference to Bechtel, the US company which currently sits at number one in the global contractor rankings and number two in the global consultants rankings. Despite its size it continues to be controlled by the Bechtel family, a fact which makes it difficult to research its financial health. It has an impressive history on the complex, civil engineering or oil and gas end of construction. Its work is well spread across the globe and overseas work accounts for up to 70% of the total, a situation which has presumably reassured many clients despite the lack of publicly available accounts. Partnerships are a strong theme in its outlook (Bechtel 2000). It works particularly closely with the US Government (see Case Study 5.15 in Chapter 5) but its current involvement in the UK illustrates that the relationships are not confined to the USA.

Comment

I have commented already in Chapter 5, section 5.2.3, about American companies and their tight relationships with customers, and Bechtel appears the ideal example of the approach. Like Bouyges, however, its scope of work is now expanding into PFI and the ownership of assets in response to current trends across the globe.

Of all the sectors, it could be argued that it is in the contractor's world that shape and scope are changing most rapidly. Major players in particular are having to change to widen their offer, and it is believed that many of the smaller players are, by contrast, narrowing their offer in ways which are similar but more exaggerated than the consultants.

To deal with this ever widening, ever more financially risky scope of work, particularly in international fields, contractors have often worked together in joint venture or through loose associations. The trend for risk-sharing appears to be growing. The following two case studies show the two ends of the risk-sharing spectrum; the first is a project joint venture, the second an industry initiative.

Case Study 8.8: The second Tagus crossing

Kvaerner Construction, part of Skanska, has a reputation for bridge building. Campenon Bernard from France has a similar reputation. When the opportunity arose for a major second crossing of the Tagus River in Lisbon, Portugal, it was always likely that these two companies would bid for the project. However, the advent of 'design, build, finance and operate' (DBFO) and its application to this project called for adjustment to both companies' approach.

Rather than competing, the two companies put together a joint venture with six Portuguese companies (NCE 1998). Kvaerner and Campenon Bernard each have a 24.8% stake in the concession company, a $1.2 bn 33 year operation for the 12 km crossing. The concession company is financed by a $320 m bank loan, shareholder equity of $27.5 m, and a European Cohesion Fund loan of $335 m for the $750 m construction cost; 2.25 bn vehicles paying tolls of $1.80 per return journey will repay the loans, shareholders and operations costs.

There was significant risk in the technically difficult cable-stayed sections of the bridge, on the land purchase for the approaches, in meeting the critical completion dates for an important Expo exhibition and in the high vehicle targets. Risk sharing through joint venture was therefore an important factor. However, the project was completed on time and to cost.

Comment

With big risks in the technology, scale of capital finances and uncertainty of revenue in one of the EU's poorer countries the lure of risk-sharing was very enticing even to two competitors with similar profiles in size and bridge specialism.

The second case study keeps the focus on Europe. It highlights the increasing influence of Europe on UK construction and the need for national contractors to seek strength in numbers in this market.

Case Study 8.9: Construction Confederation (King 1998)

The Construction Confederation of the UK is a grouping of the major UK contractors. Much of its previous work has been devoted to lobbying UK government to ensure contract conditions and public sector procurement were fair and managed to reflect the needs of the industry.

In 1998 it unveiled a new strand of work when it ran a round-table meeting in Brussels aimed at lobbying the European Commission to allow the private sector to take the lead on the Trans-European Network Schemes (TENS). TENS is a system of 12 trans-European motorway and railway transport links, seen as vital to the development of a pan-European market. The scale of the projects was such that finance for the scheme had been problematic if it remained in the public sector. Although much of the work itself would take place outside of the UK the effects of both the procurement regime, the possible work and the completed projects were likely to affect the UK market.

Comment

An interesting attempt to push a crucial part of the European market in the direction where UK contractors were rapidly building up expertise and where the client could be offered clear advantages.

How is the contractor sector different from the other sectors within construction? The non-service, non-manufacture explanation above is a simplification but nevertheless provides much of the answer and affects each part of the developing, operating and profiting aspects of the business. Business development is unlikely to show much difference, although operations will have a different emphasis, since both fixed and variable costs are important and both influence the choice of market entry into new international markets. Interestingly, the case studies bring a measure of both success and failure.

Risk *versus* reward for contractors is the most interesting aspect of the business, since the adage 'low risk, low reward' does not appear to be a good rule of thumb. Developed, developing and emerging all provide more high risk than low risk, even if profit levels might vary (as discussed in Chapter 1).

Problem solving exercises

(1) Contractors and consultants are being urged by many experts to pursue a more international workload. For companies with no previous experience this is a daunting task and involves much preparation of many different facets of information. There must be a commercial advantage for a company which is hoping to expand into new markets.

What are the commercial implications of Table 10.4 on design fee percentages in Chapter 10 for an international contractor? How could they use this information to their advantage and what risks need to be highlighted? Use examples from Table 10.4.

(2) Define the differences in approach between building material producers, consultants and contractors in seeking overseas work. Why do the differences occur?

(3) You are a contractor specialising in international projects on water supply systems. Develop a check-list of possible general

construction risks for the project outlined below so that you can fully prepare yourself for a pre-tender site visit. (We will assume that we can ignore the commercial, country and any specialist technical risks at this stage.)

The project

A notice inviting prequalification of contractors has been issued by the Russian Ministry of Construction. Contractors are asked to consider the installation of a replacement water supply system for the city of Nishni Novgorod. The Global Development Bank has provided a US$9 m loan to cover the cost of the work which will include:

- Supply and installation of all pipelines through
    - replacement of 10 km of 1000 mm diameter water supply pipeline
    - replacement of 15 km of 350 mm diameter water distribution pipelines
- Construction of a pumping station
- Reservoirs
- Administration building
- Supply, installation and commissioning of all M&E equipment

A feasibility study has already been completed and the project has been progressed as a result of this. Prequalification will be to World Bank guidelines, and these are available in either English or Russian upon submission of a non-refundable fee.

9 Project Funding

9.1 Introduction

Construction projects need funding and a client or sponsor. We have looked broadly at the types of client in Chapters 1 and 2. In Chapter 5 the traditional sources of funding for international construction were reviewed, with a combination of aid, public and private sector in poorer countries through to increasing dependence on private sector funding as the country becomes more developed. This is a very simplistic explanation of a complex system which is always in the process of change. However, this system will be used in explaining the broad concepts of how a country funds its construction and infrastructure development but with the addition of the concept of 'design, build, finance and operate' (DBFO) into the diagram in Fig. 9.1.

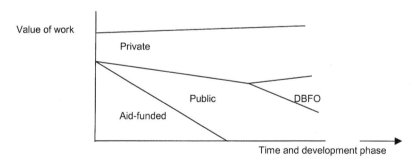

Fig. 9.1 Funding sources.

There are thus four forms of project funding to explain: aid, public sector, private sector and DBFO. These broad categories mask a whole range of alternatives, many of which are hybrids which incorporate some of two or three of the forms. They are not confined to the construction sector; many of the methods employed have been developed for other more sophisticated sectors and adapted to construction.

In theory, a client organisation starting with a blank piece of paper and an idea requiring construction of major physical infrastructure would have the option to choose any of these forms. The choice would then be on the basis of best value and most appropriate quality. In practice, client organisations are constrained by legislation, availability of finance, their own stance on risk and risk-sharing and time or project cycle timetables, so that the choice is very often limited to one or, at best, two of the options.

9.2   Pure public or private sector funding

Traditional public sector funding or private sector funding of projects remains the major source of construction work across the world. These are the projects where it appears that the project is being funded by a single source such as a government department, a major company or a local government agency. Major projects, in particular, are seldom funded by a single source although for most purposes they will appear to be so. The single source models which dominate most domestic markets are based on the idea that the client organisation will pay for the construction and maintenance of the infrastructure, as the cost arises. There is, however, huge variation of how they work both within and across countries.

Public sector projects funded in this manner have included much of the Hong Kong Airport project. The major risks of paying up-front for construction and relying on construction companies to complete on time lie with the public sector, a significant problem for all, from the poorest to the most developed. Such risks are a drain on the limited resource of the public sector purse of the host nation and consequently there has been considerable movement towards models which share the risk with the private sector through DBFO, or which tap the capital markets using publicly traded bonds (Pierson 1995). The graph in Fig. 9.2 is adapted from a presentation by NatWest Markets (1996) showing the range of options available.

The funding of projects by the private sector has likewise seen change which pushes it towards greater sharing of risk and a smoother payment profile; forms of procurement which are similar to DBFO in form or leasing are all employed across the globe.

Thus the traditional public sector clients will continue to be important major players in the sector although the sources and vehicles by which funding is put in place will widen in scope.

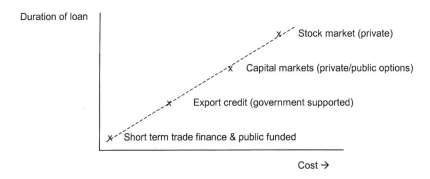

Fig. 9.2   Duration *versus* source (Adapted from NatWest Markets (1996)).

9.3   Aid funding

Morris *et al.* (1995) reported that the total value of aid pumped into global projects in 1994 was in excess of $100 bn. If all of this was construction-related work it would have represented about 3% of the global construction market, and in times past a large proportion of it would have been directed towards construction and infrastructure development projects. However, the aid sector has seen rapid change in recent times and construction or major infrastructure-related work is a much smaller proportion of the work (the same report splits the World Bank aid budget of $22 bn into a number of sectors, of which construction-related work appears confined to transport, energy and water supply and sewerage, representing 33% of the total).

In broad terms aid contracts are split into three forms: supply contracts, works contracts and technical assistance/service contracts. Aid can take various forms to match the need and the source: long term loans, loans or grants for technical assistance, grants for emergency relief work, grants in support of loans or other funding or guarantees to cover payments or a mix of the above.

There are a large number of governmental and non-governmental agencies which specialise in acting in a sense as the client or sponsor body. However, the source of the money which becomes aid is generally the OECD's Development Assistance Committee – the world's richest 22 countries plus the European Union (OECD 2000). Charities have also become major players in disbursing both their own funds and some of the government contributions.

Aid contribution is split between bilateral funding, where the donor nation decides itself where the aid will be used, and multilateral funding, where it is pooled with the contributions of others for the use of organisations such as the World Bank.

Requests for aid assistance originate from recipient countries. The recipients are grouped into two: category one which is the developing nations (in our terminology) and category two the poorest (the emerging), listed in Chapter 5.

For multilateral aid loans the stages of the project cycle are strictly observed with check-points throughout the identification, preparation, appraisal, negotiation, implementation and evaluation process. Importantly the work is put out to open tender at most of these stages, although there can be a local content clause inserted. The host or recipient nation acts as the client body with the donor nation or multilateral agency acting as the lender of the funds.

Bilateral loans are often tied or restricted to the recipient agreeing to use contractors or service providers from the donor nation. Confusion (and plain deviousness) has played a part in aid distribution. The agenda of donor and recipient nations can be very different, with the donors often wishing to see their own companies benefit from the work created by the funding and the recipients seeking the lowest possible price. There is, however, plenty of scope for recipients to play various donors off against each other to seek the best deal, and the large number of channels through which aid can be directed affects transparency.

In order to prevent a 'credit race' between competing donor nations seeking maximum influence and export creation, the member states of the OECD agreed guidelines in 1976, called the OECD consensus. This sought to fix maximum credit periods and minimum interest rates for loans. These were augmented by the Helsinki package of 1991 which set further guidance on early consultation and further restricted the use of grant and loan packages to the poorest recipient nations.

In essence, many of the rules are aimed at loans rather than non-refundable grants. Grants are a small proportion of the aid disbursement (as an example the Asian Development Bank disbursed $733 m in grants and $43 180 m in loans between 1966 and 1993. Grants represent 1.6% of the total (Morris *et al.* 1995)).

Multilateral loans are disbursed by development banks, of which the World Bank is the largest (a budget of $22 bn in 1992 (Morris *et al.* 1995)). The World Bank has other methods of raising finance and has structured its disbursement in levels of need:

donations from donors feed concessionary loans for the poorest or emerging nations, loans raised through the sale of bonds feed slightly less concessionary loans to developing nations and the bank now provides equity leverage and guarantees to investors.

Bilateral loans are disbursed through the export credit agencies of each of the donor nations. The agencies can finance a maximum 85% of a total contract price and operate medium term loans – 8.5 years for developing nations or 10 year periods for emerging (the poorest). The loans are at preferential fixed (discounted) rates, with the rates under constant review by OECD. The rates are typically based on government bond yields and are below commercial rates, although they vary across donor nations. A sample of rates from 1995 (Morris *et al.* 1995) shows how the rates vary across nations in line with government bond yields which affects competitiveness (see Table 9.1). The export credit agency contracts a bank to provide the loan, and funds the difference between commercial rates and the discounted rate.

Table 9.1   Export credit agencies' interest rates from 1995.

UK sterling	9.25%
US dollar	7.70%
Deutschmark	7.11%
French franc	8.38%
Japanese yen	3.60%

Loans and grants can be tied together but there are rules since the net effect of this is that the grant further reduces the discount cost of the loan (see note on Helsinki above). The loans arranged through an export credit agency serve the recipient nation with a cheap loan but they also serve an equally important function for the donor nation – the loan has 'developed' world status in terms of the security of payment, thus leaving the donor nation's companies less exposed to the extra risk associated with a poor country's loans, although at the taxpayer's expense.

There are other forms of Government support linked to project funding which add to the securitisation and protection for the exporter from the donor nation. The tied nature of much of the bilateral aid restricts competition. But more importantly, export credit agencies also provide guarantees and insurances which cover risks such as expropriation, war, restrictions on payment and other specific political risks (ECGD 1993). These are risks which would not be covered by normal insurance cover sought through the private sector.

Case Study 9.1: Pergau Dam (Idiculla *et al*. 1997)

We have already looked briefly at this project in Chapter 8. It is worth focusing however on the financing arrangements for the project. The 1991 agreement between the UK Government, Malaysian Government, the contractors and the client called for a capped project value of $621 m, of which $459 m was provided by the UK Government and the difference was provided by the Malaysian Government. The UK portion was set up as an ECGD loan to cover 100% of any costs of UK or European sourced exports and to be repaid over 8.5 years under the terms of an initial grace period of 5.5 years and interest charged at 0.081% per annum.

Comment

The conditions of this loan are so clearly close to an interest-free loan that the project became a celebrated case which led to the tightening of international rules covering the linking of grant and loans. An aid grant package in this case had been used to pay for the interest.

9.4 Design, build, finance and operate

First of all, it is useful to establish what exactly is 'design, build, finance and operate' (DBFO), or its alternative forms. The common terms for work which has been extended well beyond the bounds of design office and construction site include:

■ DBFO – Design, build, finance and operate
■ BOO – Build, own and operate
■ BOOT – Build, own, operate and transfer
■ PFI – Private finance initiative
■ DCMF – Design, construct, maintain and finance

The above are all variations on a theme – the situation where one party provides the services for design, build, finance and operate. There are, of course, significant variations between some of the above and they reflect the evolving practice in this field and the differing desires of client governments in procuring infrastructure projects through a long-term or even permanent concession. The basic premise in procurement of this type is that the developer is responsible for the project from arranging finance through to long-term operation and maintenance.

Most construction companies have already had experience of the extension of responsibility in moving from the traditional split-up of work into design and build packages. With DBFO the construction company therefore looks at further extension of responsibilities whereby it includes finance and operation in its offer. Of the three sectors that we have studied to date it is the contractors who have been most affected by this new type of procurement route although, increasingly, consultants are also being drawn into the finance and operation aspects of the business.

It has been argued that international contractors working in developing and emerging markets have already experienced the financial extension of responsibilities through their involvement in 'project finance' types of activities. Raising finance for commercial ventures is a typical way of doing business and it is an important part of international corporate life. However, the concept of 'one project' finance has, in the past, been handled through arrangements with governments: either a sovereign guarantee from the host national government or a financial package backed by the contractor's home government.

Although viewed as a new phenomenon this form of financing an activity has a long history (Freshfields 1996). Studies have shown similar forms of work in France in both the sixteenth and nineteenth centuries. The unique features and, by implication, the critical factors for success or failure lie in the extension of responsibility attached to the added activity. Returning to our basic equation of risk *versus* reward, it would be true to say that both are magnified as a result of the larger roles. It is therefore critical that risk is properly assessed, reward is well forecast and securitisation (through insurance, contract arrangements, etc.) is tight and correctly focused. As a result, paperwork has increased substantially and the early projects of this form were heavily criticised for being overly bureaucratic.

The time-scales, and more importantly, costs involved in operation and securing finance have led to financiers and lawyers playing a large part in the development of projects of this type. Their approach to risk and risk assessment has permeated this type of approach as all parties strive to limit their liability. Often projects are described by financiers and lawyers and the best references are in journals or publications from these sources. However, it is the concession companies, typically construction contractors (and consultants), who bear the major risks.

The finance itself is usually raised before the project is established and the money is usually a mix of equity and borrowing. A typical mix is 20% equity (i.e. shareholders) and 80% borrowed from a bank or consortium of banks. All of the funding is raised against the expectation of revenue generation from the operation of the infrastructure.

Explanation of the risks varies depending on the reference (Spence 1997). Banks, for example, concentrate on cash-flow, cost overruns and revenue. Lawyers typically concentrate on the use of contracts to push risk around, from government through the concession company and on to specialists who believe that they can better manage the risks.

Anecdotal evidence suggests that upwards of 400 separate risks have been identified in the early PFI projects, requiring contract or securitisation arrangements for each. The scale of the risk is illustrated by further anecdotal evidence that major UK DBFO projects involve, on average, $1.5 m of pre-tender work. This itself is a major risk since pre-tender work is no guarantee of work. Pre-tender work is, however, critical since it is at this point of the project that risk allocation is decided.

The major categories of risks are generally viewed as follows:

Technology risk

Often concession company risk – the risk involved in unproven technology. Many of these projects need to be leading edge to be more attractive to the government than a more traditional approach. This is a risk that the government would not take on its own projects, and lenders are also reluctant to share risk of this type.

Construction risk

Concession company risk but passed on to a contractor – viewed by many as the biggest risk since construction is notorious for cost overruns. As a result, completion guarantees and fixed price fixed time contracts are often put into place.

Operating risk

Concession company risk but passed on to a facility manager. Experience and financial stability are viewed as critical in the selection since the contract is long-term.

Case Study 9.2: The Tagus Bridge

The Tagus Bridge in Portugal was the subject of an earlier case study (8.8). It is worth noting that part of the financing involved a subsidy payment, after angry motorists blocked the bridge when tolls were raised, forcing the government to order the company to reduce the tolls (Bliss 1997).

Comment

Although the level of funding remained the same the source of revenue changed, and the operating risk exposed adds to the insecurity for the companies involved.

Supply risk

Concession company risk contractually passed to suppliers, but where the project requires a steady supply of resource (e.g. a power plant requiring fuel), the supply, its quality and its price are usually tied into a contractual agreement.

Offtake (revenue) risk

Often a concession company risk but requires negotiation with government where revenue is particularly risky. The concession company will mirror its supply contracts with offtake contracts where possible and will negotiate with government on risky projects (such as toll roads) looking for additional security through payment on basis of availability rather than use.

Case Study 9.3: The M1–M15 project

The M1–M15 motorway project in Hungary, due to complete a missing link in motorway between Budapest and Vienna, has had a troubled start to its operational life. Traffic volumes have been reported at 30% to 50% less than forecasted, leading to the need for restructuring. The problems have been attributed to high toll charges rather than lack of demand (Private Finance International 1997).

Financial risk

Often the concession company risk but requires negotiation with financiers. The sheer size, scale and long-term nature of the financial modelling creates risk for all of the financial players and the debt-equity ratio usually becomes a reflection of the size of the risk viewed by the financiers (since lenders will force the concession company to raise its equity stake).

Environmental risk

Liabilities for potential environmental problems remain a question mark.

Political risk

Ideally the government risk but often they try to push it elsewhere. The risks of *force majeure* or changes of tax regime, legislation or regulation which adversely affect the project can only partially be met through insurance if the host government is unwilling to accept these risks.

Case Study 9.4: Thailand's reputation

It has been reported that toll road projects in Thailand are avoided by international financiers because of the political risks arising from one project (Private Finance International 1997). The $250 m Bangkok Expressway (to build an expressway through Bangkok) was funded by an international consortium and built by the concessionaire, a Japanese–Thai Joint Venture. After a dispute between the partners, a court ruling transferred the responsibilities to the Thai partner, forcing the Japanese partner to sell its stake to a Thai company.

Legal risk

The cross jurisdictional nature of the whole complex operation (e.g. coal from Indonesia to feed a power plant in India run by an American company) requires a strong view to be taken on contracts and the validity of contracts in such complex situations.

Case Study 9.5:   Asetco

A construction project to build a polyethylene plant in Stavropol, Russia was financed by an export credit backed loan (Private Finance International 1997). During the construction period the country moved from the communist system to a free market system, sales contracts were in fact marketing agent contracts, state guarantees were not enforceable since the Ministry of Chemical Industries did not survive the changes and the plant was sold off to new owners, who promptly disappeared from view.

Comment

The risks are numerous and the horror stories suggest that caution is indeed required.

9.5   The markets for DBFO opportunities

The market for DBFO is clearly global. It is useful to look at some examples by way of explanation of the type and scope of opportunity available.

The UK is viewed as a relatively mature market and is frequently quoted as an example of excellence. The UK government's Treasury Department keeps track of projects and publishes lists of completed projects together with examples or case studies of individual projects.

Case Study 9.6: PFI in the UK

By April 1997 (HM Treasury 1998) PFI projects covered all manner of infrastructure from the Channel Tunnel rail link, through to the Skye Bridge, prisons, hospitals and office buildings. A total of over 60 projects were viewed as 'signed deals'. The vast majority of these involved construction of an asset followed by operation. However, some of the projects have extended beyond construction to, for example, the provision of a computer system for existing facilities.

One example, the Lewisham extension to the Docklands Light Railway, has been outlined by the Treasury. The capital cost of $300 m for the 25 year concession was funded through a $240 m bank borrowing with the rest in equity funding and public sector contributions. The PFI contractor was a consortium of John Mowlem (UK contractor), Hyder (UK consultant and infrastructure owner), London Electricity (UK infrastructure owner) and Mitsui (Japanese trading house).

Comment

Viewed as the market leader, the UK has spawned opportunity for many of the UK players to export their services and experience and for foreign players such as Bouygues to enter the market. It has thus become a very international market.

The second case study moves to a specific type of market.

Case Study 9.7: The electricity sector

A good global example of DFBO at work is the electricity sector (*Economist* 1998). Originally a public sector dominated industry the sector has, since the 1970s, quickly transformed itself to seek much of its new investment through DBFO-type concessions. In the developed world, North America, Japan, the EU and New Zealand have all made substantial plans to liberalise the public monopolies and then move into private sector concessions. In the developing world, South East Asia, India, Pakistan and much of South America and Eastern Europe have aggressively sought privatised solutions for new electricity generation.

The sector itself presents some interesting problems. Transmission of electricity is a natural monopoly since the owner of the wires will hold the key to business. However, the generation and delivery of electricity to homes (the two ends of the transmission system) have possibilities for competition. Generation, for example, has seen efficiency gains which make smaller plants more efficient, leaving much more scope for competition.

Comment

As a sector this is viewed as the most advanced in the development of this type of scheme. Many of the key advisers to client bodies have built up experience through this sector, as the actual contract and project arrangements vary little on many schemes. Figure 9.3 shows the key players who are involved in the development of a privately financed concession company.

Of these players the key players are the operators – critical since the concessions are often of a 20 to 25 year period – and the concession company: contractors and sometimes consultants often fill roles as concessionaire, contractor, operator and shareholder. The equity or shareholders will typically pay for 20% to 25% of the cost, with the rest coming from bank loans or grants.

The ultimate client bodies are in general public sector, although often they will not take full ownership until after the concession period. There are alternative forms of DBFO as the variations in section 9.4 suggest. Some systems incorporate leasing arrange-

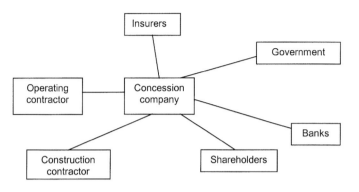

Fig. 9.3   Stakeholders in DBFO.

ments (rather than actual ownership being transferred for a limited period), and some make no inclusion for the transfer back to the public sector.

Case Study 9.8: STAR in Malaysia (Yin 1994)

From an initial list of three proposals the STAR consortium was chosen for the 60 year concession to run the first Kuala Lumpur light rail transit. The project called for construction of a 12 km track at a total cost of M$ 1.25 bn. The STAR consortium included Taylor Woodrow (UK) and AEG Westinghouse (USA) with a combined share of 45%, together with a group of Malaysian pension fund organisations. Funding was through bank loans of M$850 m, export credit and equity.

Taylor Woodrow were then contracted to build the tracks and stations while AEG supplied the rolling stock and signal equipment. The track was led along a disused railway track, avoiding the risk of acquiring private land. The foreign exchange risks on equipment import, a major item in projects of this type, were borne by a Malaysian bank, which then hedged the risk in financial markets.

The other politically sensitive risk of fare-setting was dealt with through agreement on an annual renegotiation and a threshold agreement. If the fares did not rise in line with agreed levels then the government would compensate the consortium. This agreement managed to attract a total of 18 banks into a syndicate for the loan portion of the financing.

Comment

The case study provides a skeleton of the typical arrangements of a project of this type. There is an on-going debate on what, how and if aid bodies should be involved in this type of arrangement. The huge risks involved are potentially catastrophic for the companies involved and securitisation or government level support is often vital.

Case Study 9.9: Paiton power project in Indonesia (*Project and Trade Finance* 1994)

The $2.6 bn 1300 MW Paiton 1 Power Station in Indonesia was a leading private finance project possibility in 1994. The agreed power purchase agreement (PPA) was based on a tariff which fluctuated for any price change, i.e. if the exchange rate or the cost of raw fuel increased then the cost to the client would be increased with the developer largely unaffected. However, the potential concessionaires were given only 6 months to negotiate all financing or they would have to renegotiate the PPA.

The international joint venture with an exclusive right to negotiate comprised Mission Energy (USA) 22.5%, Mitsui (Japan) 22.5%, GE Power Funding (USA) 20% and Batu Hitam (Indonesia) 20%. $600 m of the total cost was to be provided through equity from the four partners; $450 m was being provided by commercial bank loans (presumably at commercial rates), and a further $1.4 bn was being provided in export credits by the Japan Export-Import Bank (JEXIM) and by the US ExIm Bank. The remainder was being provided through a complicated political risk insurance.

Comment

The equity-debt ratio of 3.2 : 1 is quite high. However, it is clear that the role of government support in projects such as this is vital if they are to go ahead, although there are moral issues of exclusive rights to bid and coverage against loss.

This sector of the market, although a small part of the total market at present, is rapidly growing and is truly global in nature. Part of the reason for this is its dependence on large funding packages, which in turn rely on the finance markets which themselves are global in nature. The implication for the rest of the chain has been to bring in an international team in many cases. This has, of course, resulted in even the lawyers and financiers developing league tables of the biggest and best (see Table 9.2).

Table 9.2   Top 10 financiers and legal advisers.

	1994 Top Global Trade Financiers (1)	1996 Top Global PFI Legal Advisers (2)
1.	Citicorp	Linklaters and Payne
2.	WestLB Group	Baker and McKenzie
3.	Mediocredito Centrale	Freshfields
4.	Banca Commerciale Italiana	Clifford Chance
5.	Chase Manhattan Bank	Allen and Overy
6.	Dresdner Bank	White and Case
7.	JP Morgan	Milbank Tweed
8.	Barclays Bank	Norton Rose
9.	Union Bank Switzerland	Minter Ellison
10.	ABN-AMRO Bank	Skadden Arps

(1)   Reprinted from *Project and Trade Finance* (1995)
(2)   Reprinted from *Privatisation International* (1996)

Problem solving exercises

(1) List the export credit agencies of the following countries: USA, Germany, Japan, Australia, Canada, Denmark, Holland, France. What is their current annual budget?

(2) DBFO projects are becoming increasingly popular in many countries. How would a major contractor with experience of this type of work approach the project outlined below, and define the differences from a more traditional design and build project.

The project

The Romanian Department of Highways has obtained funding from the Global Development Bank to proceed with a contract seeking contractors who could upgrade and subsequently toll the existing 100 km long Bucharest Ring Road Extension. The total estimated cost of the project would be US$100 m.

10 Hints and Signposts

Having worked our way through consideration of what are the markets, who are the players and how does it all work we now turn to the process of simplifying decision-making. Like many other sectors much of the work of senior managers in international construction revolves around snap decisions. To help in making snap decisions managers often refer to a rough set of rules of engagement; what can we expect next, what workload can we expect from an opportunity, what are the basic costs and who can we expect to deal with. Very little of this is scientific but it is useful to have the background knowledge.

10.1 Winning work

It is useful to review the procedures involved in the development of an international construction project and the stages accepted in the construction cycle. The big difference between work in international construction and the typical approach to domestic construction is the need to develop a background knowledge of the full construction cycle. The five key steps, identified in earlier chapters, in World Bank projects (Spier 1985) are tracked by a number of publications:

- Identification of a country's economic needs, project feasibility study
- Loan negotiation
- Detailed design
- Project implementation
- Project completion and evaluation

Work which is not funded by multilateral agencies can and often does miss some of these stages, moving quicker than this formal process. At the other extreme, there are projects where the time-scale can be stretched to ludicrous proportions and where the project cycle is stretching over decades, crossing over the generations of experience in the same company.

Case Study 10.1: Jamuna Bridge, Bangladesh – World Bank Project (Tappin *et al.* 1998; Parker 1998)

The Jamuna River is a major river crossing Bangladesh, ranked as the world's fourth largest river. Bangladesh itself is a poor country, with much of the land forming a floodplain for the Jamuna and other rivers. Thus, any bridging of the river poses major economic and technical problems. Serious studies for a crossing were first developed in 1964, and again in 1971. Further studies in the mid 1970s indicated a cost of $1 bn and a construction duration of 13 years for a bridge link.

In 1983 a UK consultancy, Rendel, Palmer and Triton (RPT), were asked to study the feasibility of a gas pipeline crossing. They concluded that, in combination with a bridge, the cost would be $420 m and the project would take four years to build.

In 1986 RPT in joint venture with Nedeco, a Dutch consultancy, were commissioned by the World Bank to complete initial feasibility studies followed by detailed economic and technical studies. In 1987 potential financing institutions were approached since it was proposed that, although initial studies were financed by World Bank and United Nations loans, the construction phase would be financed by a wider group including Asian Development Bank, World Bank and Overseas Economic Co-operation Fund (OECF). In 1991 it was decided that the bridge was feasible and prequalification was started.

In 1994 Nedeco and RPT were appointed by the Bangladeshi government to oversee the construction works. The works themselves were split into four different packages: bridge and approach viaducts, river works and reclamation, east road approach and west road approach. After one false start, the four tenders were awarded in late 1994 at a total cost of $579 m. The main contractors were Korean, a Dutch-Belgian joint venture and a local company. The bridge was opened in June 1998.

Comment

Obviously the point at which any particular company becomes interested in a project depends on its specialist field of work. The possibility of winning the work is preceded by the need to prequalify, and beyond the skills base and the attendant reputation it is important to have networks of strategic contacts. Multilateral funded projects, such as by the World Banks, are run on the basis of open tender. This makes for a very competitive situation, and winning bids require a strong grasp of price, skills and all the information available on the project.

The World Bank itself provides numerous guides to help companies prequalify for their work, although the Bank stresses that it is the funder and not the client who makes the decision. Thus, a company wishing to bid needs to persuade officials in the recipient country, potential partners and World Bank officials of its ability to complete the work.

10.2 Market sizes

Multilateral funded work often sits in a world of its own with its own rules and players. Beyond this type of work the world of international construction consists of a set of markets, some of which are more open than others. In Chapter 2 it was suggested that a huge amount of information is needed to gauge how best to approach each market. The fundamental information required has been outlined in both text and examples in previous chapters. However, there is one important area which has been hinted at but largely left untouched – the rule of thumb.

It is always helpful to have a set of benchmarks and indicators which allow the budding manager to interpret the situation he is presented with in a foreign land with a new or potential project. The famous rule of thumb, and the numerous indicators derived from it, can serve the function of acting as a useful check throughout the project cycle. However, the more rigorous exponents, particularly academics, are very dismissive of such analytical tools, because of the gross generalisations involved. Some of the indicators have been covered in previous chapters but it is useful to summarise them, as follows.

References such as Davis Langdon and Everest (1995) provide a useful description of global markets on a national basis, providing an estimate of the size of the market and its relative breakdown. An important factor to remember is that 'big' statistics such as these are often very poor estimates of an industry which is poorly tracked.

Table 10.1 is a reminder of the rule of thumb that construction often represents roughly 10% of a country's total GDP. Relating the sector back to a well-known if often poorly measured statistic such

Table 10.1 The size of the construction market as a % of the national GDP.

	% of GDP	Construction output per capita ('000 ECU)
France	9.3	1.7
Germany	13.4	2.4
Italy	10.1	1.5
UK	7.3	1.0
Russia	6.0	0.3
Turkey	14.0	0.7

Reprinted from Davis Langdon and Everest (1995) *European Construction Costs Handbook.*

as GDP is a useful starting point in an industry where there is frequently a dearth of information. This 10% figure, of course, varies considerably depending on the country's position in the development cycle, its position within its own national business cycle and the government's attitude to management of the business cycle, as discussed in Chapter 5.

It was noted in Chapter 4 that caution is needed with pricing comparisons. However, the information itself is a useful stepping stone and Davis Langdon and Everest together with other references such as the Little Black Book (Franklin and Andrews 1996), Blue Book (Contractors Register Inc 2000) and others provide valuable information (Table 10.2). Remember that this information changes rapidly with time.

Table 10.2 Locational cost factors (UK = 100) (Franklin and Andrews 1996).

UK	100	Denmark	122
Switzerland	143	France	113
Germany	124	Austria	125
Eire	110	Holland	111

The split of work within a country also reflects the needs and is a useful pointer to the type of work available and long-term needs. A country spending a large percentage of GDP on basic civil engineering is likely to be a developing nation or a developed nation rebuilding. The long-term process, outlined by Bon and Crosthwaite (2000), of moving to developed status often manifests itself through a hotel building stage (to attract global executives), followed by business infrastructure, higher percentages of privately funded work and then repair and maintenance becoming important.

We have already discussed the quick and rough methods of assessing a nation's size and health and the point of its economic cycle through the use of surveys conducted by the likes of IMD, BERI and the *Economist* (see Chapter 5). How do these translate to ratios for construction?

Case Study 10.2: Building services in Spain from a rule of thumb analysis

Any sector within the construction industry will represent a proportion of the total construction market. For example, Davis Langdon and Everest (1995) suggest the Irish construction market represented 12% of its GDP in 1993, and that this was split into 43% residential, 33% non-residential and 23% civil engineering (all rounded figures).

In the absence of more specific information (which would be required before any firm action was taken) a first guess of the building services market in Spain for an interested company could be taken from this rough type of estimation. We can use another reference (CIRIA 1996), which suggests that building services represent typically 28% to 40% (say 34%) of building projects.

Thus Spain's market may be:

$$39 \text{ bn ECU (GDP)} \times 12\% \times 33\% \times 34\% = 0.52 \text{ bn ECU} = 520 \text{ m ECU}$$

Looking at the fragmented nature of the industry across the globe it is unlikely that any one firm would win much more than 5% of that market, i.e. 26 m ECU. This is a sizeable amount of work but may represent too small a market for many major players and would stop them from considering the market.

Comment

The rule of thumb can be useful but it must be used with extreme care. The base statistics are dubious and the combination of sources adds to problems. Moreover, at the end of the day, one major project in a small market with a large amount of building services (a major hospital development for example) would skew the figures and render them useless for a particular period. However, in the absence of anything better, they can serve as a first approximation.

More specialist fields, such as power station development, will need to follow global trends and national development cycles rather than the methods above. After this, the percentage break-down can be applied, e.g. a boiler specialist will know for example that boilers may account for around 15% of the average power station cost.

10.3 Corporate ratios

It is at the business or corporate level that the concept of rough and ready rules of thumb has seen most progress. In Chapter 3 it was noted how Tarmac, when measured against other competitors in their field, came under pressure to change direction. Much of the analysis behind the pressure came through application of rough rules of thumb about sales per employee, profit margins, return on capital employed, debt to equity ratios, etc.

Ratio analysis

The MBA school of thought dictates that management is a com-plex business and that there are ways of simplifying it for the middle or senior manager who has to keep his eye on a large number of variables. There are a number of models which com-bine a series of ratios and figures in one framework which is supposed to define the relative health of a part or whole business. There are many good references on this type of work but the stu-dent is referred to a limited list (Kay 1995; Kotler 1997) as a starting point. Many rely totally on financial ratios (Atkinson *et al.* 1986) although this is now acknowledged as being too narrow to be reliable.

How do ratio analyses relate to construction? The example of Tarmac shows this as an area of growing awareness to construction and international construction companies. The best references refer to the need for these frameworks to be adjusted to suit par-ticular markets, an area which is still under-researched for con-struction. Thus, the ratios vary across national construction markets and with time.

Set-up costs

One of the most important initial costs in developing an international business is the cost of setting up a presence or an office. This can vary depending on location, country and nature of the office (i.e. from being open full-time to a mail-box type arrangement). However, it appears that serious players typically make allowance for an expatriate type of arrangement where, initially, they fly in a trusted member of staff who sets up the office and runs it for a number of years.

Case Study 10.3: Setting up in Chile

Using the expatriate arrangement as a base case, *Building* (1995) have produced typical figures for setting up an office in Chile (see Table 10.3). The article notes that a similar exercise for Italy would produce similar figures and the author's experience of South East Asia would suggest similar figures for that location.

Table 10.3   Design fees and other costs.

Salaries	
project engineer	$60–100,000 a year
project director	$100–200,000 a year
secretary	$1,000–2,000 a month
On-costs	1.4 × net pay
Housing	
two bed flat	$1,000–2,500 a month
family house	$4,000–5,000 a month
Education	
Private	$1,000 a month
Office rents	$20/m^2 a month

Obviously the figures quoted in Table 10.3 are slightly out of date. A useful source of information is the *Economist* which provides up-to-date cost of living indices.

This type of arrangement is slowly being replaced by a local staff set-up or a short-term expatriate replaced by local staff. This trend has two effects: it replaces an expensive expatriate with cheaper local staff and it puts pressure on expatriate packages so that the old figures quoted above may not be quite so wrong. Potential expatriates need to be aware of their value and again there are numerous sources of information such as the *Economist* or *Building* magazine (Cavill 1999).

Profits

Profits are an interesting but usually confidential aspect of project costs. Despite this, it is possible from company annual reports to develop a reasonable estimate of the level of profit involved, which presumably is derived from projects.

ENR (1995a) refers to gross profit for contractors in the region of 2.5% to 5% with net profit generally less than 1%. This is very much in line with the performance of contractors in the UK. Since international work is viewed as more risky, presumably most companies would look to the top end of the range suggested. Of course in situations where there is greater leeway there is always the desire to look for higher profits.

Consultants are even harder to pin down on profit. Again the few references suggest similar figures to the contractors. Building material producers are, however, in a different league as the analyses in Chapters 3 and 6 show.

10.4   Project costs

The various core components of a project and their associated costs are well tracked by quantity surveyors, as noted in section 10.2. While quantity surveyors remain the main source of expertise for this type of work the average non-QS member of staff is well advised to have a grasp of general break-down of costs and the sources of this information. International work, as explained earlier, often requires staff to be aware of a little bit of everything.

Design fees and costs

These are a small but important aspect of construction projects. There are various studies to show the comparative costs of design fees in different countries. Table 10.4 is taken from one source

Table 10.4 % of construction value.

	Architect	Structural	M&E	QS	Other	Total
Belgium	8.00	0.50	1.00	0.75	0.25	10.50
France	4.85	2.50	—	0.65	0.90	8.90
UK	5.50	2.75	3.25	3.25	—	14.75
USA	6.20	—	—	0.30	—	6.50
Japan	4.00	1.35	2.20	1.35	—	8.90
New Zealand	4.50	2.10	2.20	2.00	—	10.80

Reprinted from ENR (1995b), Source: Hanscomb/Means Report

(ENR 1995b) which provides a split for the variety of professional consultants involved. The costs, taken from work restricted to developed countries, vary significantly. Reasons for this are numerous: wage levels in each of the countries, the scope of work and the methods employed.

Accuracy of estimates

The accuracy of estimates is another area which requires some thought before embarking on projects. Again variation is enormous but in domestic situations in, for example, the UK the contract will often protect both parties when variation outside of an agreed norm occurs. This reliance on contract and contract protection has less security in international markets since culture and social environments can be different.

A UK survey (Lenssen 1996) suggested that contractors can estimate typically to within 3% to 15% of the true cost. It is an area of huge risk, and contingency therefore, whether through alleviation measures or financial lump sums within the bid, is even more critical in international work. The same survey suggested consultant's costings are even less accurate. It is believed, however, that consultants have more scope to trim their costs to meet estimates during the work. Again, however, international conditions restrict the scope for flexibility.

Design and construction

Design and construction projects require up-front work before a bid can be submitted. Although design can represent about 2.5% to 10% of the total cost when finished, at the up-front stage the percentage can be much higher. Anecdotal evidence suggests that a reasonable estimate is 15% at that stage.

Agents' fees

Agents' fees are often a hidden but substantial cost. They are to be avoided where possible since their introduction is often associated with corruption (see Chapter 5). A report (ENR 1995c) on the subject suggests that, for example, Middle East markets require 5% to 10% fees.

10.5   Future issues

There are many studies which address the question of future trends, such as the UK's Exportism project (Thorn *et al.* 1997), Japanese forward planning (RICE 1994) or Malaysia 2020 (Hakim 1999) and these can be interesting and useful. Up to this point, however, this book has deliberately avoided speculation about future prospects either in terms of markets or sectors. The intention has been to present tools and information and let readers use them to make decisions for themselves. It is, however, possibly appropriate to look at some of the trends which have established themselves in construction or related sectors, and which may have a strong effect on the future behaviour of international construction players.

Information technology

The IT revolution has had a significant effect on many companies. Chapter 2 highlights the value of the internet in searching for information. The author has found it invaluable in the completion of this work, since even obscure, incomplete projects such as Bakun Dam and Camisea (see Chapter 5) are well covered by articles on the internet. Chapter 7 highlighted consultants' use of IT to reduce cost and time of delivery by using world-wide networks of offices to share out work, sometimes using time zones to ensure 24 hour working. In this respect, document management has become a key aspect of the benefits of IT.

Thus, there is a solid foundation of work practices incorporating IT which acts to reinforce companies' ability to work internationally. It has, however, only scratched the surface of what is possible and research is currently pushing the barriers further. Procurement and tendering through IT has become a hot topic (Guss 1996; Technopolis 1999; Economist 2000) although there have been legal and system compatibility problems in earlier experiments.

GIS, or geospatial data management with the logging of data linked to points on a mapping system, is becoming an increasingly important service available to construction companies (NCE 1998a) so that they can record their work. Again, IT has proved invaluable in the development of these techniques.

However, it is communication improvement through IT which is heralded as the area promising most for international construction. Document transfer (as discussed above), e-mail, video conferencing and virtual offices (Guss 1996) all have the potential to improve communication. In the rush to promote IT, however, many appear to forget that communication starts with an element of trust, and face-to-face communication remains invaluable.

Sustainability

This has become an overused word taken to mean all manner of different things. In this text, we shall stick to official definitions and the two important factors:

(1) Raising awareness of intergenerational effects of development
(2) Raising awareness of social, economic and environmental effects of work

This still allows many queries around how this is balanced in a rational, beneficial way. However, the arguments often boil down to balancing the still important economic effect of new development with resource scarcity, pollution control and ethics.

The subject has often appeared to incorporate the arguments of many pressure groups. Indeed, the reason why projects such as Bakun or Camisea are widely reported on the internet is because of groups opposed to such development. In the case of Bakun, there are issues revolving around the destruction of valuable (many would say scarce) jungle for a project which was widely believed to be marginally economic. From a social point of view, jungle tribes will be moved, raising further ethical considerations.

With Camisea there are similar arguments, together with a fear of possible pollution from the development of an oil source and pipeline across a fairly hostile environment. In both cases, it must be remembered that the location is a developing nation where economic growth has become a political priority. Thus, there is or may be an issue of different cultural aspirations from developed world priorities, and resolving the issue is seldom as simple as it first appears.

It is clear, however, that resource scarcity and pollution control or reduction will become an important factor for construction in the future. Already there have been many articles on the pressure on water resources (Mylius 2000) and how it will shape future priorities, while new materials are scrutinised in ever more detail to assess their contribution to improving pollution and reducing resource scarcity.

Japanese R&D

Perhaps not uppermost in many players' minds outside of Japan this is an interesting development which may yet feed into international construction. The sheer scale of funding which has gone into construction R&D in Japan has led to an enormous output, although much of it remains uneconomical for widespread use.

A 1995 report (Nahapiet 1995) highlighted the fact that over $2 bn is spent annually on construction research and development. The 40 largest contractors each had their own research institute, with the top five spending an average 1% of their turnover directly on research. Recession has undoubtedly reduced the enthusiasm for R&D since then, although work continues at a reduced scale (Nakayama 1999).

Japanese R&D has tended to concentrate on five key areas in recent years, as follows.

High-rise buildings

All of the major contractors have produced proposals for very tall buildings or sky-cities as they are often termed. The motivation for this comes from the Japanese belief that their island has limited space for development, although this has global resonance in the above sustainable development debate. Much of the work is coordinated by Japan's Ministry of Construction (NCE 1998b) which is striving to develop the concept of a 1000 m hypertower for Tokyo.

Underground space

Again the motivation comes from pressure for development space. This has led to significant R&D into tunnelling, car parking, storage and living in an underground environment (Penta Ocean 1991). It has been reported (Nahapiet 1995) that the Ministry of Industry and Trade (MITI) has a dedicated programme on the potential of deep underground space, with an annual budget of over $15 m.

Site process improvement

The most visible aspect is in robotics where the Japanese have long been acknowledged as the leaders in this technology (Nahapiet 1995).

Fig. 10.1

Fig. 10.2

Kawasaki Artificial Island sits in Tokyo Bay, Japan, and served as the starting point for tunnel boring for the Tokyo Bay Aqua Line. On completion of boring it is now used as a ventilation shaft. (Photos courtesy of Penta Ocean Construction Ltd, Tokyo, Japan)

Earthquake

This more immediate problem arises from Japan's earthquake prone position and its continued memory of past failings and disasters. Again, significant amounts of time, money and effort are devoted to R&D in this area.

Environment

This is the more recent of the key areas and appears to divide into two: work on improving the natural environment through pollution reduction, resource use improvement etc. (Penta Ocean 1991), and the provision of 'comfortable living' (Nahapiet 1995) by beautifying and enhancing the environment. Little of this work has been seen beyond Japan, but it is very probable that it will begin to filter through in future years.

10.6   Round-up

In this book we have looked at international construction in the context of (1) the global economy and (2) construction as an industry with sub-sectors. It appears to be more complex than the domestic scene but with the same basic structures and services delivered.

The case studies used are designed to reinforce special characteristics that arise through different cultures, different client needs or through the needs of the different sub-sectors. DBFO has been briefly examined since it may be the way of the future for many construction projects. Trends, patterns and rules of thumb have been an important part of simplifying the picture and the analysis likewise has sought to structure and clarify rather than quantify.

The ultimate intention has been to make the reader aware of the scale of internationalisation of the construction industry. A recent survey of building sites in London (Cavill 1998), where work regulations are generally restrictive to immigrant workers, revealed site staff from Germany, Italy, the Czech Republic, Romania, Lithuania, Australia, New Zealand, USA, South Korea and South Africa. All of these people are coping with the social and cultural pressures of not only working abroad but also living abroad, and I hope that the reader now has a taste of that.

Problem solving exercises

(1) A medium sized Western European contractor wishes to expand beyond its domestic market. It has specialised design and build expertise in immersed tube tunnel technology and has been presented with details of three opportunities in recent weeks. The projects, all of a similar size, are in Russia, Hungary and Portugal.

The company's directors believe that they have the in-house capability to assess the technical risks and the project's commercial risks but are unsure of in-country risks, threats and other opportunities. You have been asked to provide an immediate overview of the markets over the next two days, followed by a full report in three weeks' time.

(a) Gather together as much information as possible to develop an overview and comparison of the three markets in order to select the most appropriate market for further study
(b) Outline further information required for the full report stressing possible sources, reliability and priorities. (See Appendix, Hints and Model Answers for Problem Solving Exercises.)

(2) Rules of thumb are a useful tool. Discuss the advantages and disadvantages of this statement and give examples to explain your conclusions.

References

References have been split by chapter to ease sourcing material for the reader. This has entailed some references being repeated in a number of chapters. The two most heavily used references have abbreviated titles: ENR is the magazine *Engineering-News Record*, published by McGraw-Hill, and NCE is the magazine *New Civil Engineer*, published by Emap.

Chapter 1

Batchelor, C. (2000) Construction: global growth in civil engineering to top 6%. *Financial Times* 11/2/00.

Bon, R. & Crosthwaite, D. (2000) *The Future of International Construction*. Thomas Telford, London, UK.

Davis Langdon & Seah (eds) (1995) *Construction and Development in Vietnam*. E & FN Spon, London, UK.

Davis Langdon & Seah (eds) (1997) *Asia Pacific Construction Costs Handbook*. E&FN Spons, London, UK.

Economist (1995) *A Survey of Multinationals* supplement 24/6/95.

Economist (1998) *East Asian Economies* supplement 7/3/98.

Economist (1999) *A Survey of the World Economy* supplement 25/9/99.

ENR (1995a) Top Global Contractors. *ENR* 28/8/95 pp.99–100.

ENR (1995b) Top 200 International Design Firms. *ENR* 24/7/95 pp.41–42

ENR (1999) Top 225 International Contractors. *ENR* 16/8/99 pp.66–67.

ENR (2000) Top 200 International Design Firms. *ENR* 17/7/00 (on-line) http://www.enr.com

Franklin and Andrews (1996) *The Little Black Book*. Franklin and Andrews, London, UK.

IMF (2000) International Monetary Fund *World Economic Outlook* (on-line) http://www.imf.org/external/pubs/ft/weo/2000/01/index.htm

Martin-Fagg, R. (1996) *Making Sense of the Economy*. Thomson Business Press, London, UK.

Mylius, A. (1998) Can the East pay off ? *NCE roads supplement* June 1998 p.21.

NCE (1995) *Consultants File*. Emap, London, UK.

NCE (1996) *Contractors File*. Emap, London, UK.

OCAJI (1991*) Japan's Construction Today*. Overseas Construction Association of Japan Incorporated, Tokyo, Japan.

OECD (2000) *The DAC List of Aid Recipients* (on-line) http://www.oecd.org/dac

REMIT (1991) *Vietnam – the next Asian Tiger?* Report prepared by consultant REMIT for British Consultants Bureau, London, UK.

Thomson, A. I. & Oakervee, D. E. (1998) *Hong Kong International Airport – Construction*. Proceedings of the Institution of Civil Engineers (Special Edition August 1998), London, UK.

Thorn, M. F. C. *et al.* (1997) *Technology Support for Civil Engineering Exports*. Institution of Civil Engineers, London, UK.

Wheeler, P. (1998) Island Race. *NCE* 25/6/98 pp.26–27.

Williams, D. (1997) Informal discussion on construction statistics. Department of the Environment, Transport and Regions, London, UK.

Chapter 2

ACE (2000) Association of Consulting Engineers home page (on-line) http://www.acenet.co.uk

Barrow, C. & Lawn, D. (1996) *Czech Republic – A Guide to the Construction Market*. Gleeds, London, UK.

BCB (2000) British Consultants Bureau home page (on-line) http://www.bcbforum.demon.co.uk

CIA (2000) Central Intelligence Agency *The World Factbook* (on-line) http://www.cia.gov

Cooper, P. (1995) Myth of overseas profit is exposed. *Construction News* 3/8/95.

CPA (2000) Construction Products Association home page (on-line) http://www.constprod.co.uk

Davis Langdon and Everest (eds) (1995) *European Construction Costs Handbook*. E & FN Spon, London, UK.

DTI (1993) *Export Directions*. Department of Trade and Industry, London, UK.

DTI (2000) *International Trade and Investment*. Department of Trade and Industry (on line) www.dti.gov.uk/intrade/index.htm

EBRD (1996) European Bank for Reconstruction and Development *Annual Report* (on-line) was available at http://www.ebrd.com/english/index.htm

Economist (1995) *Internet Supplement* 1/7/95.

Economist (1997a) *Big Mac index* 12/4/97.

Economist (1997b) *Market Indicators* 17/5/97 p.156.

EFCA (2000) European Federation of Engineering Consultancy Associations home page (on-line) http://www.efcanet.org

EGCI (2000) Export Group for Constructional Industries home page (on-line) http://www.egci/co.uk

EIB (1996) European Investment Bank *Annual Report* (on-line) was available at http://www.eib.org/pub/pub.htm

Holley, A. (1995) *In Depth Country Appraisal of Czech Republic.* Building Services Research & Information Association, London, UK.

IMD (1996) *The World Competitiveness Yearbook.* Institute for Management Development, Lausanne, Switzerland.

IMF (2000) International Monetary Fund World Economic Outlook (on-line)   http://www.imf.org/external/pubs/ft/weo/2000/01/index.htm

UK Embassy (1997) *Czech Economic Roundup.* March 1997. UK Embassy, Prague, Czech Republic.

US Department of Commerce (1997) *Market Access and Compliance* (on-line) was available at http://www.itaiep.doc.gov

Chapter 3

Atkinson, L. *et al.* (1986) *The Manager's Handbook.* Sphere, London, UK.

Barrie, G. & Billingham, E. (1996) Culture clash unseats Myers. *Building* 3/5/96 p.7.

Brandenburger, A. M. & Nalebuff, B. J. (1996) *Co-opetition.* Doubleday, USA.

CIRIA (1994) *Control of Risk.* Special Publication no. 125. Construction Industry Research and Information Association, London, UK.

Economist (1997) Why too many mergers miss the mark. *Economist* 4/1/97 pp.59–60.

Economist (2000) Fading fads. *Economist* 22/4/00 pp.72–73.

Ernst and Young (1994) *A Guide to Producing a Business Plan.* Ernst and Young, London, UK.

Freshfields (1996) *International Construction Contracting.* Freshfields, London, UK.

Henley Management College (1995) *Creating your own Business Plan.* MBA Course Notes, Henley Management College, UK.

Kay, J. (1995) *Foundations of Corporate Success.* Oxford University Press, Oxford, UK.

Kotler, P. (1997) *Marketing Management*, 9th edn. Prentice Hall, New Jersey, USA.

Mawhinney, M. (1997) *A Strategic Review of Tarmac PLC*. Assignment report for Henley Management College, UK.

Mawhinney, M. (1999) *Virtual Client Users Manual*. Assignment report for Henley Management College, UK.

National Power (1997) *Through Life Project Management Risk Seminar*. Notes received at a seminar on risk management, October 1997, London.

Porter, M.E. (1980) *Competitive Strategy*. Free Press, USA.

Shaw, B. (1996) The external environment. In *Foundations of Management* MBA course notes. Henley Management College, UK.

Chapter 4

Berry, L.M. & Houston, J.P. (1993) *Psychology at Work*. WCB Brown & Benchmark, Los Angeles, USA.

Brandenburger, A.M. & Nalebuff, B.J. (1996) *Co-opetition*. Doubleday, USA.

Connaughton, J.N. (1996) *Value by Competition – A Guide to the Competitive Procurement of Consultancy Services in Construction*. SP117, Construction Industry Research Information Association, London, UK.

Davis Langdon and Everest (eds) (1995) *European Construction Costs Handbook*. E & FN Spon, London, UK.

ECI (1997) *East meets West – The Challenge of Globalization*. Proceedings of April 1997 Berlin Conference. C008-1, European Construction Institute, Loughborough, UK.

Economist (1994) Bribonomics. *Economist* 19/3/94 p.96.

Economist (1997) Who will listen to Mr Clean? *Economist* 2/8/97 p.58.

Economist (2000) Gifts with strings attached. *Economist* 17/6/00 p.22.

Freshfields (1996) *International Construction Contracting*. Freshfields, London, UK.

Haralambos, M. & Holborn, M. (1991) *Sociology Themes and Perspectives*, 3rd edn. Harper Collins, London, UK.

Hawkesworth, R. (1994) *Business Environment*. BA European Studies course notes, University of Wolverhampton, UK.

Jones, M. (1998) Short pile fraud hits Hong Kong central. *NCE* 4/6/98 p.3.

McWilliam, F. & O'Reilly, S. (2000) How to overcome cultural differences. *NCE* 9/3/00 p.22.

Meikle, J. *et al.* (1998) *International Construction Cost Comparisons*. Results of project completed for Department of Environment, Transport and Regions, London, UK.

NCE (2000) What the papers say. *NCE* 17/2/00 p.9.

Pietroforte, R. (1996) *Building International Construction Alliances*. E & FN Spon, London.

Soong, K. K. (2000) *Never-ending folly of Bakun* (on-line) http://www.rengah.c20.org/news/20000616.htm

Stanton, M. (1996) Organisation and human resource management: the European perspective. In *Strategies for Human Resource Management* (ed. M. Armstong). Kogan Page, London, UK.

Takaiwa, K. (1985) The relationship between man-hours and machine-hours in field construction work in overseas projects. *Bulletin of Japan Society for Mechanical Engineering*, 28 (243) Sept. 1985.

Then, S. (1996) *All systems go for Bakun Dam* (on-line) http://st-www.cs.uiuc.edu/~chai/berita/960430-960530/0543.html

TradePort (2000) *Foreign Trade Barriers – Malaysia* (on-line) http://www.tradeport.org/ts/countries/malaysia/tbar.html

Transparency International (1999) *Corruption Perceptions Index* (on-line) http://www.transparency.de/documents/cpi/index.htm

Ungoed-Thomas, J. (2000) UK companies in £1.2 m bribes trial. *Sunday Times* 11/6/00 p.2.

Upstream (1999) Camisea man at top table. *Upstream* 8/10/99.

Waboso, D. (1996) Exercising the body of knowledge. *NCE* 18/4/96 p.28.

Chapter 5

Asahi Evening News (1989a) Progress in Construction Mart. *Asahi Evening News* 19/4/89 p.5.

Asahi Evening News (1989b) Unfairness is Misunderstanding. *Asahi Evening News* 13/5/89 p.5.

BERI (1996) Business Environment Risk Intelligence *Profit Opportunity Recommendation*. Private investor service report now available on line at http://www.beri.com

Bolton, A. (1998) Consultants reel as Asian meltdown hits jobs. *NCE* 15/1/98 p.4.

Bon, R. & Crosthwaite, D. (2000) *The Future of International Construction*. Thomas Telford, London, UK.

Building (1996) Top 500 Europe's leading Contractors and Materials Producers. *Building* supplement 6/12/96.

Building (1998) Top 500 Europe's leading Contractors and Materials Producers. *Building* supplement Dec.98.

Building (1999) Taking on the world alone. *Building* 27/8/99 p.3.

Building and Construction News (1995) US Dismay over Japan's closed door policies. *Building and Construction News* 23/8/95 p.1.

Construction Europe (1995) Japanese Rig Construction Bids. *Construction Europe*, May, p.7.

Davis Langdon Consultancy (1996) *Review of Construction Markets in Europe*. March. Davis Langdon Consultancy, London, UK.

Davis Langdon and Everest (eds) (1995) *European Construction Costs Handbook*. E & FN Spon, London, UK.

Dilley, P. & Taga, K. (1995) Kansai International Airport terminal. *Civil Engineer International*, Jan/Feb.

DoE (1996) Department of the Environment *Digest of Data for the Construction Industry*. HMSO, London.

Economist (1996) Country risk ratings. *Economist* 6/1/96 p.100.

Economist (1997) Business environment. *Economist* 17/5/97 p.156.

Economist (1999) A Survey of the Twentieth Century. *Economist* 11/9/99 pp.38–40.

Economist (2000) Mori's B Team. *Economist* 8/7/00 pp.83–4.

EIC (1999) European International Contractors overseas contracts statistics (on-line) http://www.eicontractors.de/vonoc.htm

ENR (1992) Commerce Department sees dim outlook for construction in 1992. *ENR* 6/1/92 p.5.

ENR (1995a) Top 200 International Design Firms. *ENR* 24/7/95 pp.41–42.

ENR (1995b) Top 200 International Contractors. *ENR* 28/8/95 pp.99–100.

ENR (1999) Top 225 International Contractors. *ENR* 16/8/99 pp.66–67.

ENR (2000a) Top 200 International Design Firms. *ENR* 17/7/00 (on-line) http://www.enr.com.

ENR (2000b) Headline news (on-line) http://www.enr.com

ENR (2000c) Top 400 Contractors (on-line) http://www.enr.com

ENR (2000d) Top 500 Design Firms. *ENR* 10/4/00 pp.92–5.

ENR (2000e) Top 10 Building material stocks (on-line) http://www.enr.com/tracker

Flanagan, R. (1998) *Technology for exports*. Lead presentation at Exportism workshop, July, Shinfield Grange, University of Reading, UK.

Fowler, C. (1997) *Building Services Engineering*. Reading Production Engineering Group, UK.

GCAO (1993) *Reference Materials – the Current Situation in Japan*. Preparatory notes for a mission to UK, General Contractors Association of Osaka, Aug. 1993.

Grant, B. (1992) Dutch fined $28m over illegal building cartel. *Construction Today*, March.

IMD (1997) *The World Competitiveness Yearbook*. Institute for Management Development, Lausanne, Switzerland.

IMF (2000) International Monetary Fund World *Economic Outlook* (on-line)   http://www.imf.org/external/pubs/ft/weo/2000/01/index.htm

Ishii, Y. (1991) Quotations. *Construction Today*, June, letters page.

Japan Times (1989) 14 Construction firms banned by Pentagon. *Japan Times* 4/10/89.

JFCC (1993) *Construction in Japan*. Japan Federation of Construction Contractors, Tokyo, Japan.

King, D. (1998) Contractors to lobby on Euro Links. *Building* 6/3/98 p.8.

Kotler, P. (1997) *Marketing Management*, 9th edn. Prentice Hall, New Jersey, USA.

Kuroki, K. (2000) Informal discussion on Japanese building materials sector, Tokyo, Japan.

Lam, P. (1991) *Report on a critical comparison of the construction procurement and contracting systems in Japan, Singapore, Malaysia and Hong Kong*. Hong Kong Polytechnic.

Lenssen, S. (1999) Mysteries of the East. *NCE* 4/3/99 p.13.

Ministry of Construction (1992) *From Obtaining Construction Business Licenses to Winning Contracts*. Research Institute of Construction and Economy, Tokyo, Japan.

Morby, A. (1996) Mason on Top of World as Amec dream is realised. *Construction News* 24/10/96 p.2.

Morris, R. *et al.* (1995) CPRC Export Finance Workshop. June, London.

Nahapiet, H. (1995) *Time for Real Improvement: Learning from Best Practice in Japanese Construction*. Chartered Institute of Building, Ascot, UK.

Nakayama, S. (1995) Informal discussion on Japanese presence in Europe, London, UK.

NCE (1998) Illegal concession closes French toll road. *NCE* 19/2/98 p.7.

NCE (1999) UK consultants quit Asian states. *NCE* 4/3/99 p.5.

NCE (2000a) Consultants forecast South East Asia recovery. *NCE* 30/3/00 p.11.

NCE (2000b) CTRL contracts put out to tender. *NCE* 4/5/00 p.9.

NCE (2000c) Straight to Sweden. *NCE* 13/7/00 pp.14–16.

NCE (2000d) What the Papers say. *NCE* 20/7/00 p.7.

NCE (2000e) Is becoming part of a global group the best way forward for UK consultants to compete on the international stage? *NCE* 13/4/00 p.20.

NCE (2000f) Funding freeze threat forces Boston Artery cost cuts. *NCE* 16/3/00 p.5.

NCE (2000g) Big dig bigger problems. *NCE* 30/3/00 pp.12–15.

NCE (2000h) Boston costs escalate further. *NCE* 13/4/00 p.5.

OCAJI (1991) *Japan's Construction Today*. Overseas Construction Association of Japan Incorporated, Tokyo, Japan.

OECD (2000) *The DAC List of Aid Recipients* (on-line) http://www.oecd.org/dac

RICE (1991) *Japanese Economy and Public Spending* Works. Research Institute of Construction and Economy, Tokyo, Japan.

RICE (1994) *Japanese Economy and Public Spending Works*. Research Institute of Construction and Economy, Tokyo, Japan.

Royse, D. (1998) Brazilian contractor identified as Bush Foundation donor. *Naples Daily News* 22/10/00.

Santero, T. & Westerlund, N. (1976) Confidence indicators and their relationship to changes in economic activity. *OECD working paper no. 170*, Paris, France.

Schwab *et al.* (eds) (1997) *The Global Competitiveness Report*. World Economic Forum, Davos, Switzerland.

Sidwell, A.C. *et al.* (1988) Japanese, Korean and US Construction Industries. *Source Document 37*, Construction Industry Institute, USA.

Thorn, M.F.C. *et al.* (1997) *Technology Support for Civil Engineering Exports*. Institution of Civil Engineers, London, UK.

US Census (2000) May 2000 *Construction at $809.3 Billion Annual Rate* (on-line) http://www.census.gov/ftp/pub/const/c30_curr.html

Wolferen, K. (1989) *The Enigma of Japanese Power*. Macmillan, London, UK.

World Bank (2000) *Country profiles* (on-line) http://www.worldbank.org

Chapter 6

Davis Langdon and Everest (2000) *A study of the UK building materials sector*. Prepared for the DETR, London, UK.

EAG (1998) Export Action Group *Annual Report for Building Materials*. DETR, London, UK.

Economist (1999) Bagged cement. *Economist* 19/6/99 p.96.

International Construction (2000) French foiled in Blue Circle bid. *International Construction*, June.

JCB (2000) Home page (on-line) http://www.jcb.com

King, D. (1998) The opportunist. In 'Top 500 Europe's Leading

Contractors and Materials Producers', *Building* supplement, pp.36–7.

Kotler, P. (1997) *Marketing Management*, 9th edn. Prentice Hall, New Jersey, USA.

Mawhinney, M. (1997a) *Germany as a building material market*. Presentation to the Lighting Federation, Milton Keynes, UK.

Mawhinney, M. (1997b) *Strategic review of Tarmac PLC*. Assignment Report for Henley Management College, UK.

Morby, A. (1997) Redland braces itself for a pitched battle. *Construction News* 23/10/97 pp.18–19.

Thompson, R. (1998a) Cat fites. *NCE* 7/5/98 pp.16–19.

Thompson, R. (1998b) Wheels of fortune. *NCE* 16/7/98 pp.14–16.

Wolf, M. (1995) Makers back in profit and confident of more to come. *Construction News* 3/8/95 p.21.

Chapter 7

Dar Al-Handasah (2000) Home page (on-line) http://www.darcairo.com

JCCA (2000) Association of Japanese Consulting Engineers home page (on-line) http://www.jcca.or.jp''

Macneil, J. (1998) Making the most of your connections. *Building* 6/3/98 pp.48–49.

Maunsell (2000) Home page (on-line) http://www.maunsell.com.

NCE (2000) *Consultants File 2000*. Emap, London, UK.

Nikken Sekkei (2000) Home page (on-line) http://www.nikkensekkei.com

Normile, D. (1992) Man made island settles in the sea. *ENR* 13/4/92 pp.22–6.

Oliver, A. (1998) Profile Neil Francis. *NCE* 26/2/98 p.14.

O'Sullivan, B. (2000) How does the relatively small Canadian structural engineer Yolles win some of the most prestigious international projects? *International Construction*, Oct 2000 (on-line) http://www.intlconstruction.com/features

Parker, D. (2000) Arup's expanding universe. In *Consultants File 2000*. Emap, London, UK.

Winney, M. (1998) A world of their own. In *Consultants File 1998*. Emap, London, UK.

Wolton, O. (2000) Going global staying local. *International Construction*, Mar 2000 (on-line) http://www.intlconstruction.com/features

Yolles (2000) Home page (on-line) http://www.yolles.com

Chapter 8

Barrie, G. (1998) Bouyges: pushing for a top 10 placing. *Building* 3/ 4/98 pp.16–17.

Bechtel (2000) About Bechtel (on-line) http://www.bechtel.com/ aboutbech

Billingham, E. (1996) Stealth warning. *Building* 25/10/96 pp.38–40.

Bouyges (2000) Home page (on-line) http://www.bouygues.fr

Cook, A. (1998) Flat top hard hats and flip flops. *Building* 27/3/98 pp.44–46.

Davis Langdon and Everest (eds) (1995) *European Construction Costs Handbook*. E & FN Spon, London, UK.

Jones, M. (1997) Pergau Pride. *NCE* 6/11/97 pp.22–5.

King, D. (1998) Contractors to lobby on Euro Links. *Building* 6/3/ 98 p.8.

NCE (1998) In under the wire. *NCE* 26/3/98 pp.14–15.

NCE (2000) *Civil Engineering Contractors File*. Emap, London, UK.

Parker, D. (1998) Warning to limpets. *NCE* 12/3/98 pp.14–16.

PSIRU (2000) *Saur and Bouygues* (on-line) http: //www.cosatu.org.za

Skanska (2000) Home page (on-line) http://www.skanska.com

Whitelaw, J. (2000) Special relationship? *NCE* 21/9/00 pp.14–15.

Chapter 9

Bliss, N. (1997) Characteristics of road infrastructure projects of interest to project financiers. In *Global Project Finance* seminar, Jan, London.

ECGD (1993) *Credit Terms*. Export Credit Guarantee Dept, London, UK.

Economist (1998) Power to the People. *Economist* 28/3/98 pp.91–2.

Freshfields (1996) Project Finance, 4th edn. Freshfields, London, UK.

HM Treasury (1998) Private Finance Initiative website (on-line) http://www.treasury-taskforce-projects.gov.uk

Idiculla *et al.* (1997) The Pergau Hydroelectric Project Part 1: project management. *Proceedings Institution of Civil Engineers Water, Maritime and Energy*, **124**, September, 139–49.

Morris, R. *et al.* (1995) CPRC Export Finance Workshop, June, London.

NatWest Markets (1996) *Forging Successful links between UK and Greek Contractors*. Seminar Nov, London.

OECD (2000) *The DAC list of aid recipients* (on-line) http: //www.oecd.org/dac

Pierson (1995) Financing projects in mature and emerging markets. *Project and Trade Finance*, March, pp.10–11.

Private Finance International (1997) *PFI 5th Anniversary Special* 16/7/97 pp.60–82.

Privatisation International (1996) Infrastructure Adviser League Table. *Privatisation International*, November.

Project and Trade Finance (1994) Quest for power. *Project and Trade Finance*, Nov, p.48.

Project and Trade Finance (1995) Trade Finance Global Arrangers. *Project and Trade Finance*, February.

Spence, G. (1997) The Financier's View: Risk allocation taken from Private Finance Initiative. *Joint UK-Japanese Working Group for Construction seminar*, Sept, London.

Yin, B.K. (1994) Star rolls out the quality BOO deal. *Asian Survey*, Oct, pp.4–5.

Chapter 10

Atkinson, L. *et al.* (1986) *The Manager's Handbook*. Sphere, London, UK.

Bon, R. & Crosthwaite, D. (2000) *The Future of International Construction*. Thomas Telford, London, UK.

Building (1995) Chile South America. *Building* 13/10/95 pp.44–45.

Cavill, N. (1998) London Calling. *Building* 24/7/98 pp.38–41.

Cavill, N. (1999) International Salary Guide. *Building* 3/9/99 pp.38–42.

CIRIA (1996) *Value Management in Construction – A client's guide*. Construction Industry Research and Information Association, London, UK.

Contractors Register Inc (2000) *The Blue Book* (on-line) http://www.enr.com/bluebook

Davis Langdon and Everest (eds) (1995) *European Construction Costs Handbook*. E & FN Spon, London, UK.

Economist (2000) Kobe's dream. *Economist* 22/7/00.

ENR (1995a) *Plastics and Composites in Construction* supplement p.3.

ENR (1995b) US lowest in design fees. *ENR* 27/3/95 p.76.

ENR (1995c) On corruption, all's relative. *ENR* 24/7/95 p.31.

Franklin and Andrews (1996) *The Little Black Book*. London, UK.

Guss, C.L. (1996) Virtual teams, project management processes and the construction industry. In *CIB Conference on Information Highway*, Bled, Slovenia, June.

Hakim, A. (1999) *Smart Schools: Can Malaysia make the Quantum Leap?* (on-line) http://www.malaysia.cnet.com/Internet/Guidebook/Smartschools

Kay, J. (1995) *Foundations of Corporate Success.* Oxford University Press, Oxford, UK.

Kotler, P. (1997) *Marketing Management,* 9th edn. Prentice Hall, New Jersey, USA.

Lenssen, S. (1996) For blindingly obvious read mostly ignored. *NCE* 9/5/96 p.15.

Mylius, A. (2000) Keeping the peace. *NCE* 20/4/00 pp.14–15.

Nahapiet, H. (1995) *Time for Real Improvement: Learning from Best Practice in Japanese Construction.* Chartered Institute of Building, Ascot, UK.

Nakayama, S. (1999) Informal discussion on state of Japanese construction market, Tokyo, Japan.

NCE (1998a) *IT Focus* supplement. Spring.

NCE (1998b) Oil rig knowhow saves space on Japanese hyper-tower. *NCE* 24/9/98 p.8.

Parker, D. (1998) Binding Jamuna's braids. *NCE* 18/6/98 pp.14–15.

Penta Ocean (1991) Penta Ocean Construction Technical Fair 1991 (in Japanese), Tokyo, Japan.

RICE (1994) *Japanese Economy and Public Spending Works.* Research Institute of Construction and Economy, Tokyo, Japan.

Spier, G.L.E. (1985) Going for World Bank business. *Civil Engineering,* Sept.

Tappin, R.G.R., van Duivendijk, J. & Haque, M. (1998) The design and construction of Jamuna Bridge. *Bangladesh Proceedings of the Institution of Civil Engineers,* November, London, UK.

Technopolis (1999) *Partners in Innovation Final Report.* Technopolis, Brighton, UK.

Thorn, M.F.C. *et al.* (1997) *Technology Support for Civil Engineering Exports.* Institution of Civil Engineers, London, UK.

Appendix
Hints and Model Solutions for
Problem Solving Exercises

There are 25 questions spread across the 10 chapters. The questions are a combination of tests of knowledge, analysis and simple maths. While there are no right or wrong solutions to any of the questions, all answers need to be based on some solid evidence.

The limited amount of information provided in each case is typical of many situations in the real world where time and lack of information are a common problem. The reader is invited to assess each situation and then develop a solution, but it is important to remember that the objective in the real world would be to fill in the knowledge gaps as quickly and efficiently as possible. The aim of the exercises is to make the reader think and explore the subject further.

Hints have already been provided with two of the questions, but the following section provides five additional model answers.

Chapter 1 Question 2 (p. 23)	A simple case study of the UK has been provided.
Chapter 3 Question 2 (p. 63)	A skeleton for a possible answer has been provided.
Chapter 3 Question 3 (p. 64)	A scoring system for the exercise has been provided.
Chapter 4 Question 3 (p. 89)	The summarized results have been provided courtesy of Davis Langdon Consultancy and DETR.
Chapter 10 Question 1 (p. 108)	One interpretation of the information has been provded.

Acknowledgement must be made of the kind help of Davis Langdon Consultancy and DETR for their permission to use information in question three in Chapter 4, and to Teesside University for use of adapted material from the International Construction module.

Chapter 1: Question 2

An overview of the UK in the late 1990s has been provided

While the UK experience is in many ways representative of a number of countries, in other ways it is almost unique in showing the influence of the international on a seemingly domestic market. In 1993 GDP in the UK stood at approximately £550 bn ($825 bn). At the same time, and in the same way, the figure for the construction sector was variously estimated at between £50–55 billion ($75–83 billion), which represents about 8% to 10% of the whole UK economy in value. This is considerable but smaller in size and therefore less influential than, for example, manufacturing or the wider service sector.

The building material production, contracting and consulting sectors in the UK exhibit very different traits which help to explain the attitude towards and development of internationalism in companies across the sectors. In the early 1990s there were broadly 15 major UK-based contractors producing £5 bn ($7.5 bn) of turnover overseas, 20 large consultants, together with numbers of small consultants, producing £1 bn ($1.5 bn) of turnover overseas, and many UK-based building material producers with a huge amount of overseas work (no total is available). In the 1995–98 period the top 10 contractors included Tarmac, AMEC, Balfour Beatty, Bovis, Kvaerner, John Laing, John Mowlem, Kier Group, Taylor Woodrow and Costain. Their overseas turnover in 1994 ranged from 18 to 62% of total turnover, with most at the lower end of this range. As international players the big UK contractors have always shown interest in overseas markets. However, in general, contractors' league tables are heavily influenced by domestic activity. Contracting is often both very project-specific and very location-dependent, and therefore it is tied to local knowledge and local presence on the ground. As such, contractors have been slower than consultants and material producers in reacting to international influences and the tables on European contractors in Chapter 5 show only one UK company in the top 10.

The top 10 at the time was dominated by domestic names, although one foreign name had emerged, Kvaerner, while Costain had Saudi Arabian and Swedish companies as majority shareholders and Tarmac's major shareholder was a Swiss bank. Below the top 10 others emerge; notably HBG, a Dutch company, and Norwest Holst, owned by a French group.

In fact at the time seven out of the top 30 companies had foreign ownership of a type (23%). This represents a sizeable portion of a market perceived to be very domestic in its outlook. Internationalisation of the market had in a sense crept up on the UK through ownership of shares rather than as a result of a change in approach. A case study in Chapter 8 makes further reference to this creeping internationalisation. Although UK contractors have not featured significantly in European or global listings, their base market has changed. Consultancy and building material sectors are much more international in outlook.

Even domestically produced league tables for consultants refer extensively to work overseas. The ENR global lists show UK consultants doing remarkably well right across the globe. They are particularly strong throughout the Commonwealth and it is often noted that UK consultants win 8% of the World Bank's available consultancy work.

By contrast the building material producers are viewed as being strong in Europe with 8 of the top 20 across Europe coming from the UK (see Table 5.5). The sector often sees its market as European, more similar in many ways to much of the manufacturing sector than to construction. This is only to be expected, since the economies of scale and market-regulatory regimes of this sector suit a European perspective. Production has become transnational in many cases and there are pan-European companies leading the way across Europe.

In stark contrast to contractors the building material producers appear to have a product which travels easily across borders. A European market has developed, and wider market blocks may exist.

Comment

Thus, there is a striking difference between different types of UK player. Consultants are often global, material producers are often European in focus, and contractors have until recently still measured themselves domestically. The above analysis, of course, relies on the evidence from the biggest players and the belief that size matters, which is a contestable assumption.

Chapter 3: Question 2

A skeleton for an answer has been provided

A market plan would need to include the following:

(1)   country/region environment – corporate needs
(2)   market environment – competitors, clients
(3)   project-specific information.

It might include reference to:

■   desk-top studies – gathering information from what sources
■   country visits
■   talking to officials, clients, competitors.

The report and plan stage would seek to establish:

■   priorities
■   specific targets
■   tactics and action steps.

The options for specific projects are likely to include:

■   supplying from elsewhere and using an agent
■   establishing a joint venture
■   setting up a manufacturing facility.

These options are subject to:

■   the degree of risk and costs acceptable
■   availability of suitable partners
■   other opportunities.

All these need to be balanced by considering the positive and negative, looking at and beyond specific projects, and reviewing the market conditions. As a very rough estimate for the project and market size:

■   project size is possibly 6% to 13% of $14 million
■   possible total construction market may be roughly 10% of GDP.

GDP and official statistics must be treated with caution. The above takes no account of finance availability (heating and ventilation equipment may be viewed as a luxury rather than a necessity with implications for financing, and there is need to forecast 3 – 5 – 10 years ahead, but with the usual caveats).

Chapter 3: Question 3

A scoring for the exercise has been provided

(Arriving at these scores involves a degree of subjectivity and therefore your score may differ from those below.)

Risk factor	Score	Note
Commercial risk		
Conditions	10	client's document
Delay damages	5	10%
Performance damages	6	10% +
Bonding	5	bond required
Fixed price	5	two-year fixed
Ground risk	2	M&E
Design responsibility	5	shared
Defects liability period	5	12 months
Exposure/Experience		
Technology risk	18	standard/complicated
Experience with partner	30	no previous experience
Geographic Risk		
Access	10	unknown – possibly jungle?
Political stability	10	difficult conditions
Language	10	non EC
Dispute recourse	10	non EC
Foreign exchange risk	6	US$ project – degree of stability
Weather	8	possible monsoon
Client	7	unknown
Project Size	40	

Score assessment	Maximum score	Actual score
Project score rating	300	192
Assessment of risk based on score	borderline medium to high	

Chapter 4: Question 3

The summarised results have been provided courtesy of Davis Langdon Consultancy and DETR

Estimate

In the absence of better information use the London model specification, the Dutch costs from Davis Langdon and Everest (1995) and adjust for inflation. A back calculation against unit cost for the industrial building provides a check. The figures below were, however, developed by Dutch specialists.

Cost in Guilders (£ in brackets)

Substructure	653,658	(202,559)	Piling as necessary
Frame	282,775	(87,628)	As UK model
Roof	286,622	(88,820)	As UK model
External walls	110,780	(34,329)	Profiled steel sheet
Windows and doors	243,422	(75,433)	As UK model (slight change)
Internal partitions	38,143	(11,820)	As UK model
Upper floor of office	23,212	(7,193)	As UK model
Floor finishes	18,555	(5,750)	Fitted carpet & linoleum
Services	534,610	(165,668)	As UK model
External works	185,917	(57,613)	As UK model (slight change)
Preliminaries	495,886	(153,668)	
Total construction costs 1997	2,873,560	(890,482)	

This does not include VAT, fees or contingency/profit.

The risks involved in completing any exercise such as this are numerous but would include differences in standards causing changes in materials and sizing, unknown local costs, currency fluctuation, the effects of time e.g. programme duration, the process of design and approvals and contract conditions.

A major risk is that it will be accepted at face value – it is a budget rather than a tender price.

Chapter 10: Question 1

One interpretation of the information required using a sample of information obtained in 1998 from the World Bank and various competitiveness indices (such as those indicated in Chapter 2)

Official headline statistics show Russia to be a huge market with a big population and thus strong long-term potential. Hungary and Portugal are smaller countries.

In terms of GDP/capita, measure of wealth (and, indirectly, funding availability) the comparative figures suggest these assessments.

Russia	= $2,400	emerging long-term potential but risky?
Hungary	= $4,200	emerging long-term potential but less risky?
Portugal	= $9,400	good medium-term potential?
UK	= $24,700	example of developed country

Foreign direct investment is one measure of confidence and success, and represents a source of opportunity for less risky projects. All three countries look broadly similar in comparison to, for example, the UK, but it must be noted that Portugal receives funding from the EU, which should be a steadying influence and source of funds and markets. GDP% change is often a measure of short-term economic health.

- Hungary 5% – best short-term
- Portugal 3%
- Russia 1% – worst short-term

Information on other statistics, for example, trade balance and current projects, would all help this picture.

Various competitiveness indices showed Russia to be a consistently poor performer, Hungary to be slightly better and Portugal the best of the three, although there were no huge differences. Items such as perceptions of financial health, infrastructure, management capability and labour quality are often measured and these showed similar trends. Summing up and choosing using the information above:

- probably Portugal looks best short-term
- Russia has the best long-term potential
- Hungary is the middle ground.

Much depends on the opportunities that exist in the country beyond the one project under consideration, and there is a need to match risk profiles with the company's needs – corporate *v.* project needs. Finally there must be a caveat on the use of such broad statistics for this exercise.

Further information required

The combined PARTS & PEST frameworks provide a useful list of information necessary. The information needs checking against possible sources, reliability of the information and a judgement on the priorities.

Index

Note: page numbers of case studies are in bold type; tables are shown in italic